MANUAL LIFTING

A Guide to the Study of Simple and Complex Lifting Tasks

Ergonomics Design and Management: Theory and Applications

Series Editor
Waldemar Karwowski
Industrial Engineering and Management Systems
University of Central Florida (UCF) – Orlando, Florida

Published Titles

Ergonomics: Foundational Principles, Applications, and Technologies
Pamela McCauley Bush

Aircraft Interior Comfort and Design
Peter Vink and Klaus Brauer

Ergonomics and Psychology: Developments in Theory and Practice
Olexiy Ya Chebykin, Gregory Z. Bedny, and Waldemar Karwowski

Ergonomics in Developing Regions: Needs and Applications
Patricia A. Scott

Handbook of Human Factors in Consumer Product Design, 2 vol. set
Waldemar Karwowski, Marcelo M. Soares, and Neville A. Stanton

Volume I: Methods and Techniques
Volume II: Uses and Applications

Human–Computer Interaction and Operators' Performance: Optimizing Work Design with Activity Theory
Gregory Z. Bedny and Waldemar Karwowski

Knowledge Service Engineering Handbook
Jussi Kantola and Waldemar Karwowski

Trust Management in Virtual Organizations: A Human Factors Perspective
Wiesław M. Grudzewski, Irena K. Hejduk, Anna Sankowska, and Monika Wańtuchowicz

Manual Lifting: A Guide to the Study of Simple and Complex Lifting Tasks
Daniela Colombiani, Enrico Ochipinti, Enrique Alvarez-Casado, and Thomas R. Waters

Forthcoming Titles

Neuroadaptive Systems: Theory and Applications
Magalena Fafrowicz, Tadeusz Marek, Waldemar Karwowski, and Dylan Schmorrow

Organizational Resource Management: Theories, Methodologies, and Applications
Jussi Kantola

MANUAL LIFTING

A Guide to the Study of Simple and Complex Lifting Tasks

Daniela Colombini
Enrico Occhipinti
Enrique Alvarez-Casado
Thomas Waters

CRC Press
Taylor & Francis Group
Boca Raton London New York

CRC Press is an imprint of the
Taylor & Francis Group, an **informa** business

CRC Press
Taylor & Francis Group
6000 Broken Sound Parkway NW, Suite 300
Boca Raton, FL 33487-2742

Printed in the United States of America on acid-free paper
Version Date: 20120522

International Standard Book Number: 978-1-4398-5663-5 (Paperback)

Library of Congress Cataloging-in-Publication Data

Manual lifting : a guide to the study of simple and complex lifting tasks / Daniela Colombini ... [et al.].
p. cm. -- (Ergonomics design and management : theory and applications)
Includes bibliographical references and index.
ISBN 978-1-4398-5663-5 (pbk.)
1. Lifting and carrying. I. Colombini, Daniela.

T55.3.L5M27 2012
620.8'2--dc23 2012015480

Visit the Taylor & Francis Web site at
http://www.taylorandfrancis.com

and the CRC Press Web site at
http://www.crcpress.com

Contents

Preface

One of our vocations as "old, hardened researchers" has always been to seek solutions for the prevention of occupational risk of biomechanical overload and to teach self-management of the problems at the source. Prevention should be done directly by those designing workplaces and jobs!

Our experience comes from constant comparisons between the need to respond to the actual needs of workers and technical staff (who require practical tools, simple and easily applicable in the field) and the need to find answers, solutions, and criteria by experimentally verified reliable methods (being often quite complex). In short, translation of the matter complexity into easily applicable prevention approaches is just one of the main goals of the Ergonomics of Posture and Movement (EPM) Research Unit, which we founded and in which we have operated for many years, under the sponsorship of Don Gnocchi, ONLUS Foundation (IRCCS Santa Maria Nascente, Milan Polo-Tecnologico).

Another important goal is to diffuse the knowledge on the matter. Following a series of meetings with colleagues and technicians, at congresses and meetings in different parts of the world, the idea to create an International School of Ergonomics of Posture and Movement was originated. Thus we created the school as an operative section of EPM (more details can be found at www.epmresearch.org). The school has its accredited teachers and has the main objective to teach different professionals, in different parts of the world and in their native language, the art of prevention according to an easy and effective approach.

In summary EPM and its schools have the following main goals in relation to the broader aim of improving health and work:

- Development of simple methods and computerised tools, for field application of research results, suitable to simplify the evaluation and management of risk by biomechanical overload
- Transformation of research results and best application practices relating to prevention and management of risk by biomechanical overload into training materials
- Development of specific training courses directed to the different professionals involved with the prevention (technical and medical professionals)
- Activation of new schools (with either private or public character) in different parts of the world (e.g., at universities, research agencies, professional interest groups, etc.), with teachers specifically trained, thus ensuring the homogeneity of the different EPM-associated schools

Following the above goals, this book responds to the application and teaching needs to make more easily measurable and therefore manageable, even by personnel not particularly skilled in ergonomics, the matter of biomechanical overload due to manual loads handling, with attention to the more complex situations. The work was carried out in close cooperation with international experts and researchers basically following existing National Institute for Occupational Safety and Health (NIOSH) guidelines, implemented to face specific aspects. To facilitate the application the approach was also implemented in Excel spreadsheets, freely available at the www.epmresearch.org website.

The International School of Ergonomics of Posture and Movement has already prepared specific training packages on the manual handling of loads in Italian, English, Spanish, and French. This book represents the main tutorial.

Acknowledgements

We thank the great team (of which only a few names are mentioned below), who cooperated with passion, love, and joy (and free of charge), contributing to produce this volume, intending to update the knowledge and offer methods to evaluate in a simple way—even very complex manual handling works—something that was previously impossible.

Thanks and nostalgic remembrance to our teacher, friend, and "father" Nino Grieco, to whom we devote the volume. Nino has taught us the beauty and joy of working for prevention, always with great attention to ethics (without great attention to ethics, it is advisable to turn to other things).

Again, thanks to all EPM friends and to the International School of Ergonomics of Posture and Movement teachers.

Many thanks also to the following friends and colleagues for their collaboration. They worked primarily in the definitions and verification of the feasibility of new evaluation criteria as well as staging of software:

Robert Fox (United States): Technical fellow in ergonomics for General Motors.

Natale Battevi (Italy): Medical doctor, specialist in occupational medicine and statistics. Senior researcher at the Ergonomics of Posture and Movement Research Unit.

Olga Menoni (Italy): Occupational therapist. Senior researcher at the Ergonomics of Posture and Movement Research Unit.

Marco Placci (Italy): Electronic and biomedical engineer. Senior consultant at the Ergonomics of Posture and Movement Research Unit.

Marco Cerbai (Italy): Safety technician and ergonomist. Senior consultant at the Ergonomics of Posture and Movement Research Unit.

Sonia Tello Sandoval (Spain): Industrial engineer, ergonomist, senior consultant in Center of Applied Ergonomics (CENEA), Barcelona.

Aquiles Hernandez Soto (Spain): Degree in kinesiology, ergonomist, director in Center of Applied Ergonomics (CENEA), Barcelona.

Daniela Colombini
The "Jefa"

Enrico Occhipinti
Director of EPM Research Unit

The Authors

Daniela Colombini has a degree in medicine with specialisation in occupational medicine and health statistics; she is a European ergonomist. Since 1985, she has been a senior researcher at the Ergonomics of Posture and Movement Research Unit–Milan, where she developed methods for the analysis, evaluation, and management of risk and damage from occupational biomechanical overload. She is co-author of the OCRA method (now standards EN 1005-5 and ISO 11228-3). She recently founded and launched as educational coordinator the EPM International Ergonomics School (which operates in different languages, such as English, Spanish, Portuguese, French, and soon German). Schools are already working with accredited native teachers in different countries, such as Spain, Chile, Argentina, Mexico, Colombia, Brazil, France, and Switzerland. She is a member of the Ergonomics Committee of UNI and represents Italy in international commissions of CEN and ISO working on biomechanics.

Enrico Occhipinti is a specialist in occupational medicine with an Ergonomics European Certificate, and is responsible for the Center for Occupational Medicine (CEMOC) working at the Department of Preventive Medicine at Foundation IRCSS Poloclinico Ca' Granda in Milan. He is professor at the School of Specialisation in Occupational Medicine, University of Milan. He is director of the Ergonomics of Posture and Movement (EPM) Research Unit, Polo Tecnologico Fondazione Don Gnocchi ONLUS–Milan and of the EPM International Ergonomics School. He has devoted more than 20 years to the study of ergonomic issues related to working postures and the prevention of work-related musculoskeletal disorders. He is coordinator of the technical committee on the prevention of musculoskeletal disorders of the International Ergonomics Association (IEA) and member of the Ergonomics Committee of UNI, and represents Italy in international commissions of CEN and ISO working on biomechanics.

Thomas Waters is the senior safety engineer at the National Institute for Occupational Safety and Health (NIOSH), Human Factors and Ergonomics Research Team. He is a certified professional ergonomist and holds advanced degrees in engineering science and biomechanics from the University of Cincinnati. As a researcher at NIOSH for the past 25 years, Dr. Waters has published more than 40 papers and chapters on manual materials handling and prevention of lower back disorders. Dr. Waters is recognised internationally for his work on the revised NIOSH lifting equation. His primary research interests include occupational biomechanics, work physiology, lower back injury prevention, and ergonomic risk assessment.

Enrique Alvarez-Casado is an industrial engineer, master of ergonomics, and master of occupational risk prevention. He is professor of ergonomics at the Department of Business Administration (DOE) at the Universitat Politècnica de Catalunya (UPC), Catalonia, and professor at the EPM International Ergonomics School. He is also a main consultant for the Center of Applied Ergonomics (CENEA), Barcelona, president of the Catalan Ergonomics Association (CATERGO), and coordinator of the work group on anthropometry and biomechanics of UNE, and represents Spain in international commissions of CEN and ISO working on biomechanics.

Important Note

The views expressed in this book are those of the authors and do not necessarily represent the views of the U.S. National Institute for Occupational Safety and Health.

1 Introduction: Aim of This Manual and Upsurge of the Problem

The rationale for this manual stems from contributions made by the Ergonomics of Posture and Movement (EPM) Research Unit toward implementing EU Directive 90/269/EEC, "Minimum Health and Safety Requirements for the Manual Handling of Loads Where There Is a Risk Particularly of Back Injuries to Workers," and transposing the directive into the national law of EU member states.

Since the transposition into the local legislations of several countries, including Italy and Spain, encompasses not only European but also international criteria, methods and tools, such as the ISO 11228 regulations or the 1993 revised NIOSH lifting equation (RNLE) [Waters et al., 1993], this manual aims to offer all those involved in workplace ergonomics, protection, and prevention some useful guidance and tools (both on- and offline). Based on the practical experience of the authors, the manual should help users apply both the international standard ISO 11228-1 [ISO, 2003] and the RNLE.

The tools were created by EPM with the support of a qualified group of co-workers in various European countries, based on extensive field experience. For over 10 years now EPM has worked with researchers at the National Institute for Occupational Safety and Health (NIOSH), primarily Thomas Waters, the principal author of the RNLE, to develop theoretical models and application tools for the study of complex (variable and sequential) manual lifting tasks. Variable and sequential tasks today represent the latest evolution of the original RNLE [Waters et al., 1993] and have become the reference method for international standards.

Since the 1970s and 1980s there has been growing evidence in the scientific literature of a link between manual materials handling (MMH), especially lifting tasks, and lumbar spine problems (e.g., lower back pain and degenerative disorders). Today, the national authorities of many countries across the globe have introduced specific standards and guidelines for performing such activities and, even more importantly, for preventing negative effects on the health of workers and thus reducing the related social and economic costs.

However, recent findings, primarily from European sources closer to the authors, would suggest that greater emphasis needs to be placed on the evaluation and prevention of manual materials handling activities. Workers are still widely exposed to such risks and disorders and injuries are still extensively reported in sectors such as agriculture, building construction, manufacturing, and health care.

Figures 1.1 and 1.2 refer to the Fourth European Working Conditions Survey (European Foundation for the Improvement of Living and Working Conditions, 2005), named the European Foundation, 2007 report [European Foundation, 2007].

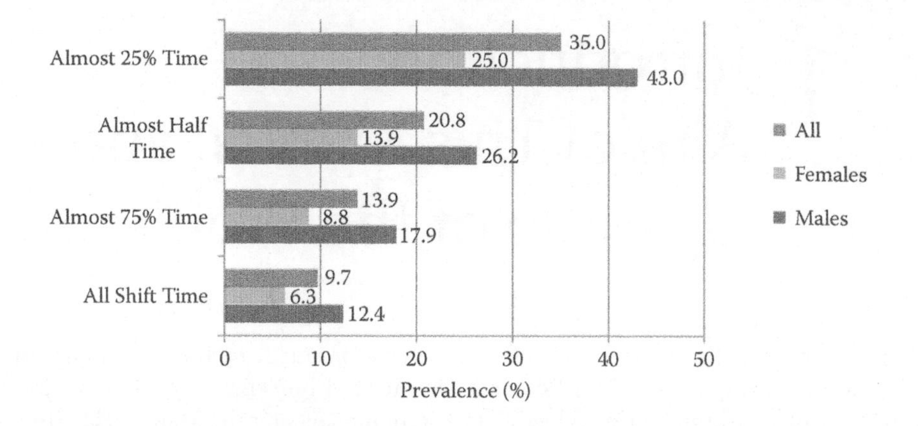

FIGURE 1.1 Prevalence of EU workers exposed to manual materials handling, by fraction of shift time and for 27 EU countries. (From European Foundation, 2007.)

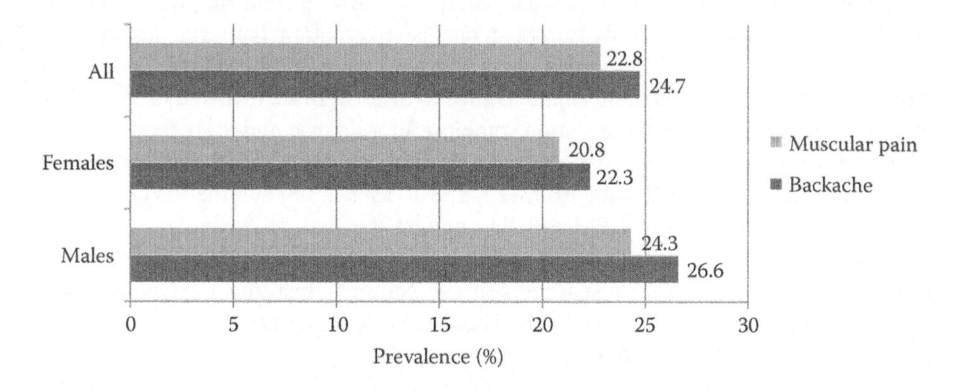

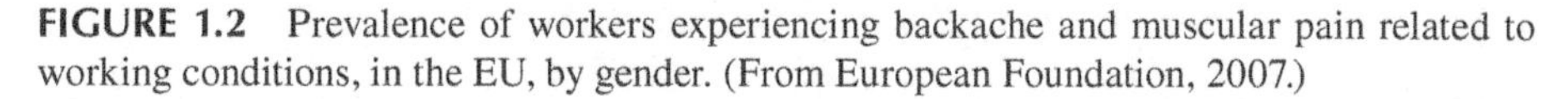

FIGURE 1.2 Prevalence of workers experiencing backache and muscular pain related to working conditions, in the EU, by gender. (From European Foundation, 2007.)

Figure 1.1 shows the percentage of workers exposed to MMH by fraction of shift time and gender. According to the survey, in the 27 countries of the European Union around 35% of workers (43% male and 25% female) spend at least one-quarter of their shift time on manual materials handling. An impressive 14% of workers (18% male and 9% female) spend 75% of shift time performing MMH tasks.

Of course the percentage of workers exposed to MMH varies depending on the sector examined: for example, for exposures of at least 25% of shift time, in agriculture and building construction, the prevalence of exposed workers (both male and female) is extremely high, at 67.1% and 64.2%, respectively; in manufacturing, the level is around 42%, while in other sectors, the level is much lower and comparable to education (11.1%), public administration (17.5%), or financial services (4%).

Figure 1.2 also comes from the European Foundation, 2007 report, this time showing the prevalence, both total and by gender, of workers in the 27 EU countries, showing work-related musculoskeletal disorders (WMSDs), i.e., backache, muscular pain of the neck, shoulders, and upper limbs. These are the most frequently reported work-related

TABLE 1.1
Reported Backache, by Gender and Weekly Working Hours

	Backache (% positive)		
Working Time	**Males**	**Females**	**Total**
10–19 h/week	14.3	12.2	12.8
20–30 h/week	18.3	20.9	20.3
31–35 h/week	16.0	18.8	17.4
36–40 h/week	23.6	22.7	23.2
41–45 h/week	25.6	22.0	24.5
>45 h/week	37.3	40.5	38.1
Total	27.0	23.6	25.6

Source: European Foundation, 2007.

disorders, with a prevalence of around 25% of workers for backache and 22–23% for muscular pain in the shoulders and upper limbs (with both disorders slightly more widespread in males). It should be noted that none of these disorders can be automatically attributed to MMH, although such tasks, together with others, may contribute toward explaining why backache ranks so highly among work-related disorders.

Again, the prevalence of workers reporting work-related low back pain differs from sector to sector, ranging from 50.5% in agriculture to 36.5% in the building construction industry and 27.5% in manufacturing, dropping to 19.5% in public administration and 16.8% in education and 12% in financial services.

Another interesting aspect of work-related backache emerges from European Foundation, 2007, i.e., prevalence of workers reporting backache in relation to weekly working time.

Table 1.1 breaks down working time by gender and number of hours per week, indicating the prevalence of work-related backache in all sectors: obviously the trend rises with the number of hours.

In viewing degenerative spinal disorders as occupational diseases related to the manual handling of loads, it should be pointed out that according to the classification of European Occupational Disease Statistics (EODS), disorders of the lower back (and neck and shoulder) region are accepted as occupational diseases by only a few member states and only for specific forms of disease. It is therefore also difficult to collect comprehensive European level data on recognised lumbar WMSDs (such as herniated disks). Conversely, musculoskeletal diseases of the upper limbs (i.e., tendonitis, epicondylitis, and carpal tunnel syndrome) are included on the EODS list and have for some time now accounted for the majority of occupational pathologies reported in Europe.

In Italy, degenerative lumbar diseases (related to manual load handling and whole-body vibration) have recently been recognised as occupational. Consequently, in recent years there has been a steep rise in the number of degenerative diseases, especially herniated disks, reported and recognised as work related.

TABLE 1.2
Prevalence (%) of Intervertebral Disk Diseases Treated as Occupational with Respect to Total Occupational Diseases Reported in the Period 2004–2008 in the Italian Manufacturing Industry

Prevalence (%) of Intervertebral Disk Diseases					
Years	2004	2005	2006	2007	2008
Prevalence (%) of Cases Treated	6.03	8.28	10.34	10.08	12.2

Source: INAIL, 2009.

Table 1.2 shows the percentage of intervertebral disk disease versus total occupational diseases reported in the period 2004–2008. The table shows how quickly herniated disks have become a significant occupational disease in Italy (especially among agricultural workers), accounting for around 12% of all WMSDs reported (or paid in damages) by the insurer. The figure is likely to continue increasing over the coming years.

On the other hand, in some countries, like Spain, Italy, or the UK, accident figures address acute episodes of musculoskeletal disorders (acute lumbago or overexertion injury), for example, occurring after lifting of heavy loads. Where this is the case, the proportion of these accidents versus the overall accident rate is high. Corresponding prevalences tend to be much higher than those for the related occupational diseases.

A high proportion of days lost in the member states of the European Union is due to musculoskeletal disorders, which have a huge impact on work-related absences. The European Commission estimates that musculoskeletal disorders account for 49.9% of all absences from work lasting 3 days or longer and for 60% of permanent work incapacity.

In some states, 40% of the costs of workers' compensation are caused by musculoskeletal diseases, and up to 1.6% of the gross domestic product (GDP) of the country itself. They reduce company profitability and add to the government's social costs.

The European Agency for Safety and Health at Work [2008] has estimated that the costs of all work-related musculoskeletal diseases account for between 0.5 and 2% of the GDP. The costs related to WMSDs have been estimated for certain European countries:

1. The Netherlands: 1.8% of GDP (1.7% for low back pain alone)
2. United Kingdom: 0.8% of GDP
3. Germany: 0.61% of GDP
4. Finland: 1% of GDP

This brief overview, along with much other relevant data that go beyond the scope of this manual and as such have not been included, strongly suggests that the problem of manual load handling and its potential health effects on working populations is still an open issue demanding the utmost attention on the part of all those involved in occupational health and prevention.

Moreover, the manufacturing industry is not the only culprit: Any area where work is largely manual, and which may present even greater risks, is of relevance, and this includes agriculture, building construction, transportation, and the service sector (i.e., commerce, utilities, health care, and welfare). Manual load handling may also take many different forms and require a wide range of analytical approaches in order to accommodate the various complexities involved.

The proposals for analysing (and preventing) WMSDs contained in this manual expand on the original RNLE method on which ISO 11228-1 [ISO, 2003] and EN 1005-2 [CEN, 2003] are both based, in order to reflect the most common real-life situations in which the manual handling of materials and loads occurs, in all their variability and complexity. However, every effort has been made to simplify the approach by using the priceless computer-intensive software applications available to all operators in the field.

A certain amount of detail is required only for specifying the organisational aspects of manual handling tasks (i.e., who, what, how, when, how much, and for how long). So far, analyses have tended to be overly generic, but now it is possible to more accurately define exposure to biomechanical overload conditions and plan more efficient technical, structural, and organisational ways of preventing risk. The proposals contained in this manual will undoubtedly help to bring about significant improvements, although there is still a very long way to go in many sectors, especially building construction. However, that subject will be dealt with in another manual.

2 A Brief History of the NIOSH Lifting Equation

2.1 POLITICAL STRATEGIES RELATED TO THE MANUAL LIFTING OF LOADS

2.1.1 Limitation of Maximum Weight Permitted

In recognition of the health consequences of repeated manual lifting on the working population, some countries have decided to use a political strategy to limit the maximum weight that a worker can lift. Of course, this is not the only factor to consider, but restricting it is at least a minimum legal safeguard.

From a historic point of view, this strategy is used mainly to protect the most vulnerable "labourer" population: children. Table 2.1 summarises data on maximum lifting weight limits by gender and age established in different countries by law or legal disposition. This information was taken from legal archives to illustrate the issue and is not meant to be exhaustive. We emphasise that some of these legal dispositions are still in force.

In 1962, the International Labour Organisation (ILO) published an informative table that fixed the maximum manual lift for occasional lifting (once per day) per gender and range of age (Table 2.2). These limits were mainly justified from the statistical study of industrial accidents and illnesses that identified manual lifting as being responsible for the increase of approximately 3 times more injury of the back, knees, and shoulders; 10 times more of the elbows; and about 5 times more of the hips.

As for the report published by ILO in 1988, 62 countries around the world established laws that limited the maximum weight permitted in manual lifting activities or in carrying of loads [ILO, 1988].

2.1.2 International Labour Organisation Convention

On 28 June 1967, the general conference of the International Labour Organisation adopted Convention 127 concerning the maximum permitted weight to be carried by one worker [ILO, 1967a]. The convention did not establish a maximum weight but did establish general principles concerning manual handling in order to guarantee the health and safety of the workers that must lift and carry loads. Citing Articles 3 and 4, it establishes that:

> No worker shall be required or permitted to engage in the manual transport of a load which, by reason of its weight, is likely to jeopardise his health or safety ... taking into account of all the conditions in which the work is to be performed.

TABLE 2.1
Maximum Lifting Weight Limits in Different Countries

			Maximum Weight (Kg)	
Country	Year of Publication	Age	Male	Female
Bolivia	—	<16	10	5
		16–20	10	10
Chile	2005	16–18	20	—
Colombia	1979	16–18	15	—
		<16	—	8
		>16	—	12.5
Denmark	2005	<18	12	12
Ecuador	—	<16	16	—
		16–18	23	—
		<18	—	9
		18–21	—	12
Egypt	—	12–15	10	7
France	1973	14–15	15	8
		16–17	20	10
Greece	—	<16	5	—
		16–18	10	—
Hungary	—	14–16	15	—
		16–18	20	15
		>18	—	20
Ireland	1972	14–16	8	8
		16–18	16	11
		>18	—	16
Italy	1934	15–17	25	15
		<15	15	5
		>17	—	20
	1967	<15	10	5
		15–18	20	15
Japan	—	16–18	30	25
Mexico	1934	<16	20	10
		>16	—	38
Poland	—	<16	16	10
	2002	16–18	20	14
Spain	1908	<16	10	10
	1957	<16	5	—
		<18	—	8
		16–18	20	—
		18–21	—	10
		>21	—	20
USSR	1921	—	—	15

TABLE 2.2
Maximum Limit Weight Suggested by ILO in Kilograms

	Maximum Weight to Lift Manually in an Occasional Way (kg)	
Age	**Men**	**Women**
14–16	14.6	9.8
16–18	18.5	11.7
18–20	22.6	13.7
20–35	24.4	14.6
35–50	20.6	12.7
>50	15.6	9.8

Source: ILO-CIS, 1962.

TABLE 2.3
Limit Weight Recommended by ILO (R128)

Age	Gender	Maximum Permissible Weight
>18	Male	55 kg
>18	Female	Substantially less than that permitted for adult male workers
<18	Male and female	Substantially less than that permitted for adult workers of the same sex; regular manual transport of loads not permitted
<16	Male and female	Manual transport of loads not permitted

And in Article 7 it cites:

> Where women and young workers are engaged in the manual transport of loads, the maximum weight of such loads shall be substantially less than that permitted for adult male workers.

The convention also adopted Recommendation 128, supplementing the Maximum Weight Convention C127 [ILO, 1967b]. The weight limitation recommended in this document is detailed in Table 2.3. It's noted that members applying this recommendation should take account of:

- Physiological characteristics, environmental conditions, and the nature of the work to be done
- Any other conditions that may influence the health and safety of the worker

In January 1987, 22 countries of the world ratified the Convention C127 [ILO, 1988]. As for the information issued by ILO [ILOLEX, 2011], 28 countries ratified it (Table 2.4).

TABLE 2.4
List of Countries That Ratified the ILO Maximum Weight Convention C127, 1962

Country	Date of Ratification
Algeria	June 6, 1969
Brazil	August 21, 1970
Bulgaria	June 21, 1978
Chile	November 3, 1972
Costa Rica	March 16, 1972
Ecuador	March 10, 1969
Spain	June 7, 1969
France	May 31, 1973
Guatemala	July 25, 1983
Hungary	January 4, 1994
India	March 26, 2010
Italy	May 5, 1971
Lebanon	June 6, 1977
Lithuania	September 26, 1994
Luxembourg	April 8, 2008
Madagascar	January 4, 1971
Malta	June 9, 1988
Nicaragua	March 1, 1976
Panama	June 19, 1970
Peru	June 19, 2008
Poland	May 2, 1973
Portugal	October 2, 1985
Bolivarian Republic of Venezuela	February 1, 1984
Republic of Moldova	December 9, 1997
Romania	October 10, 1975
Thailand	February 26, 1969
Tunisia	April 14, 1970
Turkey	November 13, 1975

It is important to note that the ratification of ILO Convention C127 did not entail the application of the maximum weight limits included in Recommendation R128. This ratification only included the risk that the manual manipulation of weight can present, also taking into consideration the issue of gender in the legal framework.

2.1.3 The European Council Directive

The European Union, with the aim to standardise the legal criterion in the different member states and also to promote the development of the legal framework, on 29 May 1990, made public Council Directive 90/269/EEC on the minimum health

TABLE 2.5
Limitations of the Maximum Weight for Gender, as a Result of the Application of the Council Directive

		Maximum Weight (kg)	
Country	Year of Publication	Male	Female
Czech Republic	2010	50	20
Denmark	2005	30	30
Estonia	2001	40	20
France	2008	55	25
Italy	2008	25	20
Poland	2002	50	20
Spain	1997	40	40
United Kingdom	1992	25	16

and safety requirements for the manual handling of loads [Council of European Communities, 1990].

Concerning the theme treated in this book, the main contribution of this council directive was to establish that whenever the need for manual lifting of loads by workers cannot be avoided, the employer shall assess the health and safety conditions of the type of work involved, taking into account factors related to characteristics of the loads, physical effort required, characteristics of the working environment, and requirements of the activity.

Some of the countries integrated this council directive in their legal framework, practically as a translation of the complete text in its official language. This was the case in Austria, Finland, Germany, Greece, Luxembourg, the Netherlands, and Sweden. On the other hand, some other countries included other technical criteria related to manual lifting risks. One of these criteria is a limit on the maximum weight to lift for male and female workers (Table 2.5).

Taking into consideration that in all the member states of the European Union the legal obligation exists to evaluate the risk of every kind of work that entails a manual manipulation of weights, one of the most important technical criteria is the direction to identify the manual manipulation of weights.

From this point, only some countries have defined legal technical criteria in order to identify the hazards of lifting weights and, as such, the need to evaluate the risk. These criteria are based upon minimum lift weights identified as potentially hazardous.

In Denmark, Spain, and Italy, the minimum weight that can be lifted that is potentially hazardous and requires its risk to be evaluated is 3 kg, in Estonia it is 5 kg, and in the United Kingdom, 3 kg for women and 5 kg for men. In general, we can affirm that only in a few countries around the world has this technical criterion been established; for example, Chile and Venezuela, 3 kg; Argentina, 2 kg.

In any case, it is clear that Europe needed to provide an evaluation procedure to define the risk for manual lifting, which must also be sufficiently versatile in order to be applied to all types of work.

2.2 THE FIRST NIOSH LIFTING EQUATION

2.2.1 Description of the Equation

In 1981, the National Institute for Occupational Safety and Health (NIOSH) recognised the growing problem of work-related back injuries and published the *Work Practices Guide for Manual Lifting* [NIOSH, 1981]. This document was created between 1979 and 1980, with the contributions of American manual lifting and back injury researchers, including Ayoub, Chaffin, Herrin, Drury, and Snook. Beginning with an exhaustive bibliographic review of epidemiological, biomechanical, physiological, and psychophysical factors, they proposed a procedure to calculate the limit weight recommended in operations of lifting of weights. In order to consider the difference between populations in the capacity to do this kind of work, they defined two kinds of limits: the maximum permissible limit and the action limit.

The maximum permissible limit (MPL) represents the value that, if it is exceeded:

- Significantly increases the rate of prevalence and seriousness of the musculoskeletal health effects
- Exceeds the biomechanical strength of compression on the intervertebral disk L5-S1 by the majority of the population, with a value higher than 650 kg; or exceeds 5 kcal/min metabolic loading
- Only 25% of men and less than 1% of women have the physical capability to work at the level of the MPL

The action limit (AL) represents the value that, if reached:

- The increase of the rate of prevalence and seriousness of the musculoskeletal health effect is low on the population exposed
- Produces a compressive force of 350 kg or 3.4 kN on the L5-S1 disk, which is tolerable for the majority of healthy workers into middle age
- Presents metabolic demands of 3.5 kcal/min or less in the majority of the working population
- Presents strength demands that accommodate more than 99% of men and more than 75% of women

The NIOSH committee defined five risk variables or factors that modify the risk condition during the lifting of weights:

- The *horizontal* and *vertical* position of hands at the beginning of the lifting (in relation with the body and the ground, respectively)
- The *vertical* displacement between the point of grabbing at the beginning and at the end of the lift
- The lifting *frequency* and the *duration* of the exposure to manual lifting tasks in the shift

The proposed formulas for the calculation of these two limits give form to the strong interrelation that exists between the different risk factors:

$$AL = LC \times HM \times VM \times DM \times FM \tag{2.1}$$

where AL is the action limit (kg), $LC = 40$ kg, and $HM = 15/H$, where H is the horizontal position of the hands at the beginning of the lift and the middle point of the ankles. In the case where the load is fragile or requires precision when placed, the longest distance measured at the origin and at the destination is taken.

$$VM = 1 - 0.004 \times \text{abs}(V - 75)$$

where V is the height of the hands at the beginning of the lift in relation with the ground. In the case where the load is fragile or requires precision when placed, the longest distance measured at the origin and at the destination of the lift is taken.

$$DM = 0.7 + 7.5/D$$

where D is the vertical difference between the point of grasping of the load at the beginning and at the end of the lifting.

$$FM = 1 - F/Fmax$$

where F is the average frequency of lifting per minute and $Fmax$ the tolerable average frequency that is obtained in Table 2.6.

$$MPL = 3 \times AL \tag{2.2}$$

where MPL is the maximum permissible limit (kg) and AL is the action limit (kg).

It is assumed that these variables have the following limits:

H is between 15 and 80 cm, considering that a load, at a distance longer than 80 cm, is increasingly difficult to grasp for segments of the population.
V is between 0 and 175 cm.
D is between 25 cm and (200 – V); for values less than 25 cm the value 25 is used.
F is between 0.2 (a lifting every 5 minutes) and $Fmax$. For frequency values less than one every 5 minutes use $F/Fmax = 0$ (or $FM = 1$).

TABLE 2.6
***Fmax* Values**

	Fmax	$V > 75$ cm	$V < 75$ cm
Duration	1 h	18	15
	8 h	15	12

2.2.2 Interpretation

Comparing the weight lifted and the calculated limits, NIOSH proposed the following interpretation in three categories:

1. Lift weight greater than *MPL*: Unacceptable. It requires an administrative and engineering control.
2. Lift weight between *AL* and *MPL*: Unacceptable for a segment of the worker population. It should be considered unacceptable without an adequate administrative and engineering control.
3. Weight less than *AL*: Acceptable. It would entail a minimum risk for most of the working population.

In the 1981 NIOSH *Work Practices Guide for Manual Lifting* (WPG), the technical document describing the NIOSH Lifting Equation, the term administrative control was defined as use of adequate selection and training for workers who are going to do this kind of work.

2.2.3 Proposals for the Analysis of Variable Tasks

In the 1981 NIOSH WPG, the authors present proposals on what criteria to apply to analyse lifting tasks that have variability in load location.

2.2.3.1 Lifting of Constant Weights Located at Variable Heights

The 1981 AL and MPL equations provided a way to assess variable geometries (vertical heights and horizontal reaches), but the method relied on an averaging process that was later shown to be inaccurate and was modified in the revised version of the equations that were developed and published in 1993 and 1994. In effect, the horizontal distances and vertical heights were averaged across the various subtasks. This did not allow the peak exposures to be captured in the assessment. The limitations of the original procedure for assessing multitask lifting jobs was discussed in the *Applications Manual for the Revised NIOSH Lifting Equation* (Waters et al., 1994; Section 2.1.2, p. 37).

2.2.3.2 Lifting of Weights and Variable Distances

A good example would be work in warehouses or goods reception, where workers handle objects with high frequency, and with different weights and shapes. In this case, each subtask represents a constant geometry (distances *H* and *V*). For this reason, in each subtask workers lift objects of different weights.

An approach is to calculate the limits of each subtask considering the partial frequency of lifting. In order to evaluate the overall task, the limits using the following variables can be calculated:

F: The total frequency of lifting.

H, *V*, and *D*: The values of the arithmetic weighted average for the contribution of the frequency of each subtask to the total frequency.

The final evaluation of the workplace will be obtained by comparing the limits obtained for each subtask with the average weight and the maximum weight lifted in each subtask, and by comparing the limits obtained in the overall task assessment with the average weight lifted considering all the subtasks.

2.3 THE REVISED NIOSH EQUATION

2.3.1 Development of the Equation

The 1981 equation was limited in terms of application, only in certain types of loads, mainly sagittal lifting tasks. Even so, it generated high interest and impact among professionals of risks prevention and labour health. For this reason, in 1985 NIOSH set up a panel of experts with the mandate to review the 1981 procedure and increase the applicability of the application to other real situations of manual lifting, which are often in workplaces. The participants in this experts' panel were M.M. Ayoub, Donald B. Chaffin, Colin G. Drury, Arun Garg, Suzanne Rodgers, Vern Putz-Anderson, and Thomas R. Waters.

In order to accomplish this, the working group created an exhaustive review of the progress made since 1981 in the scientific literature on the physiological, biomechanical, psychophysical, and epidemiological aspects of the manual lifting of loads. The result of this review was a document published in 1991 [National Technical Information Service, 1991]. During that time, they presented a revision of the NIOSH equation of 1981 to a conference in Michigan [Putz-Anderson and Waters, 1991], which is known as the revised NIOSH equation.

In 1993 this new equation was published in a scientific journal [Waters et al., 1993], where they described the rationale and justification of the selected criteria and the equation's values. Finally, in 1994, with a more didactic aim, the NIOSH published the applications manual [Waters et al., 1994], explaining how to apply the new equation to the solving, step by step, of different cases of load lifting tasks.

2.3.2 Upgrading in the Criteria

Due to the fact that each kind of lifting task imposes different biomechanical and physiological demands on the workers the developers of the Revised NIOSH Lifting Equation used different criteria to build the equation. The biomechanical aspects prevail predominantly in nonfrequent lifting tasks, the physiological criteria limit the metabolic stress and the fatigue in repetitive lifting tasks, and the psychophysical criteria limit the workload based on the psychophysically determined perception of the capacity to lift loads in all the tasks with a frequency less than 6 lifts/min and at different vertical height ranges of the lift.

2.3.2.1 Biomechanical Criteria

The biomechanical criteria of the equation of 1981 of maximum compression strength of the vertebra L5-S1 equal to 3.4 kN are retained in the 1991 revised lift equation.

TABLE 2.7
Limit Values of the Energetic Consummation Considered in the Revised NIOSH Equation

Kcal/min	$V < 75$ cm	$V > 75$ cm
≤1 h	4.7	3.3
1–2 h	3.7	2.7
2–8 h	3.1	2.2

2.3.2.2 Physiological Criteria

Compared with the equation of 1981, the value of baseline aerobic capacity was reduced to 10.5 kcal/min. In order to calculate the reduction of this limit, the committee considered the difference between the lifting that involves the entire body (the grasping height of the load in relation to the ground is less than 75 cm) and lifting that involves mainly the upper extremities ($V > 75$ cm.). For $V < 75$ cm the limit was fixed at 9.5 kcal/min, and for $V > 75$ cm they applied an additional reduction of 30%.

Moreover, they considered the influence of the lifting task's duration during the workday, fixing the following limits:

Duration equal to or less than 1 h: Limit value equal to 50% of 9.5 kcal/min.
Duration between 1 and 2 h: Limit value equal to 40% of 9.5 kcal/min.
Duration between 2 and 8 h: Limit value equal to 33% of 9.5 kcal/min.

The metabolic limits provided for in the equation, according to the height (V) and the duration (D), are given in Table 2.7.

2.3.2.3 Psychophysical Criteria

The psychophysical criteria selected for the equation are based upon the capacity of 99% of the male working population and 75% of the female labour population (roughly equal to 90% of the population given a composition of 50% males and 50% females).

2.3.3 Revision of the Equation

The main differences of the revised NIOSH equation of 1991 in relation to the 1981 equation follow:

- The minimum value of the horizontal distance H was changed from 15 cm to 25 cm, reflecting the minimum distance required to prevent interference with the front of the body.
- The load constant in the equation was reduced from 40 kg to 23 kg, due to an increase of the minimum horizontal distance and to the upgrading of the maximum limit of 75% of the labouring women. This is very similar to the weight limit established by ILO in 1962 (Table 2.8), 23 kg, in ideal conditions when all the multiplier factors are equal to 1.

TABLE 2.8
Comparison of Two Sets of Multiplier Factors in the Original Equation (1981) and the Revised Equation (1991)

	1981 Equation	1991 Equation
LC	40 kg	23 kg
HM	$15/H$	$25/H$
VM	$1 - 0.004 \times \text{abs}(V - 75)$	$1 - 0.003 \times \text{abs}(V - 75)$
DM	$0.7 + 7.5/D$	$0.82 + 4.5/D$
FM	$1 - F/Fmax$	Table 4.4
AM	—	$1 - 0.0032 \times A$
CM	—	1, 0.95, or 0.9

- The value of the frequency multiplier, instead of being calculated through a formula, is obtained by a table, depending on the duration of the task, the lifting frequency, and the grasping height. The frequency multiplier table was generated using both psychophysical and physiological criteria, as described in Waters et al. 1993.
- This combination of information was mainly obtained thanks to the psychophysical criteria for frequencies of less than 4 lifts/min, and from the physiological criteria for frequencies of more than 4 lifts/min.
- The vertical multiplier was changed in order to reduce the recommended weight by 22.5% when the load is at the ground level ($V = 0$ cm) or the height of the shoulder ($V = 150$ cm), instead of the reduction of 30% of the original equation.
- The minimum reduction on the recommended weight of the distance multiplier was adjusted from 26% in the original equation to 15% in the revised equation.
- Two new multiplier factors were introduced to take account of the influence of lifting loads offset from the sagittal plane (asymmetric multiplier (AM)) and the quality of the coupling (coupling multiplier (CM)) in the maximum acceptable capacity of lifting loads.

In Table 2.8 the main differences between the equation of 1981 and the revised equation of 1991 are summarised.

In the revised lifting equation the formula for the calculation of the recommended weight limit (RWL), with all the multiplier factors, is given as follows:

$$RWL = LC \times HM \times VM \times DM \times AM \times FM \times CM \tag{2.3}$$

For interpretation of the result to apply this equation, the committee defined a risk index, called lifting index (LI).

$$LI = L/RWL \tag{2.4}$$

where L (load weight) is the weight of the lifted object.

The design intent is that for values of $LI \leq 1$, 99% of the male working population and 75% of the female working population are protected.

For $LI > 1.0$, NIOSH stated that the task poses an increased risk for lifting-related low back pain for some fraction of the workforce. NIOSH also stated that some members of the committee who helped develop the revised equation believe that worker selection based on research studies, empirical observations, or theoretical considerations can accurately identify workers who can perform lifting tasks with an $LI > 1.0$ without an increased risk of work-related injury. These members agree, however, that many workers will be at elevated risk if the $LI > 3.0$ [Waters et al., 1994].

2.3.4 Proposal for Analysing Composite Tasks

The procedure proposed in 1981 to analyse multitask manual lifting jobs was based on arithmetical weight averages and could underestimate the risk in many cases [Waters, 1991]. In the revised equation a new approach was adopted that involves the calculation of the composite lifting index (CLI) in order to estimate the composite risk of a multiple lifting task [Waters et al., 1994].

This new method is based on the following assumptions:

- The performance of multiple lifting task increases the physical and metabolic requirement.
- The degree of increase depends upon the features of the component tasks performed.
- The increase in physical demands due to the execution of one or more lifting tasks is independent of the risk associated with the tasks already done.

This procedure is explained in detail in Chapter 6.

2.4 ANALYSIS OF THE EQUATION AND NEW PROPOSALS

2.4.1 Validation of the Revised NIOSH Equation

The publication of the revised NIOSH lifting equation had great influence on ergonomics and health and safety practitioners involved with manual handling injury prevention and on ergonomics researchers. Since its publication, various studies on the equation and its use have been conducted. Regarding validation of the equation, various results have been found. Considered here are three main levels of validation (from the lowest to the highest degree of validation) and their applications to the method of calculation of lifting index (LI) and the composite lifting index (CLI).

2.4.1.1 Validation of the Criteria

This approach of risk evaluation provides a solid base in the scientific literature for the criteria used in the equation and in the procedure to measure and evaluate each one of the risk factors and its utility for task design and intervention.

References addressing the NIOSH criteria include the following:

Analysis of the criteria used [Jäger and Luttmann, 1999] [Elfeituri and Taboun, 2002]
Analysis of the multiplier factors, their applicability, and their evaluation:
- Asymmetry factor [Dempsey and Fathallah, 1999] [Lavender et al., 2009]
- Horizontal distance [Potvin and Bent, 1997]
- Coupling factor [Honsa et al., 1998]

Different factors and their interactions [Maiti and Bagchi, 2006]

The studies largely confirm the suitability of the equation factors, but some have noted limitations in the equation as well, especially for highly variable lifting tasks. Some analyses of these problems were addressed by [Mital and Ramakrishnan, 1999] and [Dempsey, 1999], and all concluded that it was necessary to find new approaches to analyse multiple-component MMH tasks.

2.4.1.2 Validation of the Applicability

Validation has the intent to show that the result obtained with the evaluation method for a particular work situation is not unduly influenced by other factors, such as variation between analysts, etc. As such, various studies addressed the precision obtained in the evaluations of risk for different evaluators [Waters et al., 1998] [Dempsey, 2002] [Dempsey et al., 2001] [Saleem et al., 2003] [Ribeiro and Remor, 2009].

Most of these studies show an acceptable validity with high precision obtained and low variability; sometimes there are some critical aspects (such as asymmetry) that should be improved. In all cases the importance of mandatory training is underlined. This means that these methods of evaluation require an active training of quality in order to ensure a correct application.

2.4.1.3 Epidemiologic Validation

The epidemiologic validation addresses the degree to which the risk index is a good indicator of the development of work-related musculoskeletal disorders, i.e. low back disorders. The association between the revised NIOSH lifting equation lift index and back injury is suggestive, but complete validation remains elusive for a host of reasons. Even so, there are some studies carried out that suggest this association.

In 1999 [Waters et al., 1999] a cross-sectional study about the 1-year prevalence of low back pain was conducted in 204 people in the exposed group and 80 people in the unexposed group. The group of exposed workers developed its activity in a total of 50 workplaces where manual lifting is one of the predominant requirements. In each workplace they calculated the LI and CLI of each lifting task, using the maximum value as the reference. The result showed a trend in association between the increase of LI and the increase of low back pain, especially for LI values in the 2–3 range. In the same year another cross-sectional study of 353 industrial workers of 48 different industries [Marras et al., 1999] was published. This study analysed the indicator capacity of LI, taking into account the variability of lifting by means of the calculation with both the average values and the maximum values, as well as the meaning of each one of the multiplier factors. The results were that LI is a moderate predictor of high-risk situations, but it loses capacity with low-risk situations. In conclusion, they affirm that "these analyses indicate that the NIOSH approaches have predictive power for identifying jobs that place workers at risk of developing low back disorders."

More recently [Boda et al., 2010] a multisite prospective cohort study was published with the goal to validate the predictability of the revised NIOSH equation of work-related low back pain (LBP) on 258 workers without any symptoms, in a total of 30 companies. The variability of exposition was treated taking into account the maximum values of LI and CLI, as they are related. The results showed a good evidence of association between the maximum values of LI and the incident cases of LBP.

The researchers from NIOSH [Waters et al., 2011] completed the previously mentioned cross-sectional study [Waters et al., 1999]. This resulted in a total sample size for the combined analysis of 677 persons (560 in the exposed group and 117 in the unexposed group). The 393 additional workers included 37 workers in the $LI = 0$, 84 in the $0 < LI \leq 1$, 107 in the $1 < LI \leq 2$, 115 in the $2 < LI \leq 3$, and 50 in the $LI > 3$ categories, respectively. Previously, the $1 < LI < 2$ category was not significantly different from the unexposed group, and only the OR for the $2 < LI < 3$ category was significantly greater than for the unexposed group. The additional power provided by adding individuals to the study across all exposure levels, however, has shown that the OR for the $1 < LI \leq 2$ category is also significantly higher than that for the unexposed group.

Despite these preliminary encouraging results, it should be underlined that complete and definitive validation studies on the association between LI levels and low back injuries and disorders are still lacking and no definite conclusions on this issue can be derived.

For many reasons, it is very difficult to construct validation studies on something such as the NIOSH lifting equation. The nature of illness/injury data, workers moving between jobs, and the impact of physical work other than lifting are some of the complications in any such study.

However, the findings from these studies support the conclusion that the lifting index can be used to identify lifting jobs with increased risk, being a useful predictor of risk of lifting-related low back pain.

Research is needed to refine and extend the application of the RNLE to encompass a wider range of lifting jobs, such as those involving one-handed lifting, lifting in combination with pushing, pulling, and carrying, lifting in less than optimal environmental conditions, and jobs with variable task characteristics.

In this perspective, considering that in many EU countries (i.e. Italy) an active and periodic health surveillance of workers exposed to manual handling is compulsory by the law as well as a correspondent risk assessment (eventually performed using the RNLE approach), in following years we intend to conduct some cross-sectional and, if possible, perspective validation study, mainly considering LI as the exposure indicator and the incidence of acute lumbago episodes as one of the main adverse health effects.

2.4.2 Other Converging Methods

During these years, different governmental and standards producing bodies proposed their own methods, as, for instance, the method TLV® for Lifting, of the American Conference of Governmental Industrial Hygienists [ACGIH, 2005] and the method Hazard Zone Jobs Checklist, of Washington State [Washington State Department of Labour and Industries, 2008].

These methods are presented for a more extensive application for no expert users, with a more schematic procedure. But many of these methods are related to the revised NIOSH equation, and they consider it the basis for their criteria and from which they created simplifications.

Another commonly used approach to the design and analysis of manual handling tasks is provided by Liberty Mutual tables [Liberty Mutual, 2004] and the adaptation made by the Health and Safety Executive (HSE) in United Kingdom for use by their labour inspectors, called MAC [Monnington et al., 2002], published as a guidance on regulations in [HSE, 2004].

These psychophysically determined data, developed by Snook and Ciriello [1991] and others, are incorporated in part in the criteria of the revised NIOSH equation.

2.4.3 International Standards Based on the Revised NIOSH Equation

Currently the world's largest developer and publisher of international standards is the International Organisation for Standardisation (ISO), an entity recognised by the United Nations and which follows the practices recommended by the World Trade Organisation. Dating from 1991 with the signature of an agreement of collaboration, known as the Vienna Agreement, both the European Committee for Standardisation (CEN) and the ISO work together in the development of regulations to avoid duplication of work.

Since 2000 the joint work of Work Group 4 "Biomechanics" of Technical Committee 122 of CEN and Work Group 4 "Human Physical Strength—Manual Handling and Force Limits" of Subcommittee 3 of the Technical Committee 159 of ISO originated the publication of several technical regulations related to the manual manipulation of loads, the series EN 1005 and the series ISO 11228, respectively.

Concurrently, in 2003 they published two consensus standards on manual lifting, EN 1005-2 and ISO 11228-1, both incorporating the revised NIOSH equation (as described in Chapter 3).

2.4.4 New Proposals

In 2007, Thomas Waters, Ming-Lun Lu, and Enrico Occhipinti developed a proposal to analyse sequential lifting tasks (i.e. the performance of different lifting tasks by a worker during a shift) [Waters et al., 2007]. Sequential lifting tasks are quite common in industry due, for example, to rotation strategies.

The sequential lifting tasks approach is explained in Chapter 9. In addition to sequential tasks, lifting tasks may still be highly variable in that workers handle many objects of different weights at different heights and horizontal distances.

These situations are also quite common, for example, in tasks related to just-in-time material delivery, warehousing, and other logistics-related material tasks.

A voluntary working group was convened with the aim to analyse these problems and develop a procedure of analysis and had its first meeting in September 2007 in Barcelona, Spain. This group was promoted by the EPM Unit Research, and formed by Thomas Waters of NIOSH and researchers of EPM Unit Research, of the Universitat Politècnica of Catalunya (UPC), and Centro de Ergonomia Aplicada (CENEA).

After 2 years of research and meetings, they defined evaluation procedures using the criteria of the revised NIOSH equation of 1991 and presented the results to the 17th Triennial Congress of the International Ergonomics Association [Waters et al., 2009] [Colombini et al., 2009].

This variable lifting task analysis approach is explained in Chapter 8.

3 International Technical Standards for the Manual Handling of Loads: ISO 11228 and EN 1005

3.1 REFERENCE METHODOLOGY AND INTRODUCTORY REMARKS

In the early 2000s, international standards organisations began developing specific regulations concerning the manual handling of loads. ISO 11228-1 [ISO, 2003] and EN 1005-2 [CEN, 2003] focus specifically on lifting tasks, which will be explored in this chapter. Various European countries, including Spain and Italy, have adopted and used these standards as application guidelines for their national legislations (based on European Directive 90/269) concerning the manual handling of loads.

Before going further into this aspect, it should be noted that both of the aforesaid standards take the revised NIOSH lifting equation (RNLE) [Waters et al., 1993, 1994] as their reference method. Although not always explicitly stated, and despite certain adaptations, there is no question that the RNLE was the inspiration for the approach shared by these two sets of standards.

The reference masses or weight constants referred to in these introductory remarks deserve particular attention, since they will help readers to better understand the approach adopted in the rest of the manual.

The original version of the revised NIOSH lifting equation (RNLE) recommends computing the recommended weight limit (RWL), and consequently the lifting index (LI), by using a single reference mass of 23 kg (51 lb) for all potential users (regardless of gender and age). This reference value has been assumed [Waters et al., 1993] to be protective for around 90% of the healthy adult working population (male and female), and more specifically, for between over 75 and 90% of the female population and 99% of the male population.

Arguably, the advantage of using a single reference mass lies in ensuring equal access to employment, but at the same time there is the issue that irrespective of the real protection level for male and female populations, it supplies a different level of protection when gender is taken as a factor.

This aspect was discussed at length when the ISO and CEN technical standards were developed: While some countries supported the use of a single reference mass, regardless of gender (and age) factors, others (mainly in central and southern Europe), due to their national legislations or cultural mindset, preferred to envisage a similar protection level (say, 90%) for the various population categories (males and females; younger and older workers) and therefore asked for reference values to be differentiated at least on the basis of gender and age.

The solution was found in both the ISO and CEN standards by offering users a range of reference values (mass, load constant) along with information concerning the presumed protection levels that the norms ensured for the different categories of healthy workers. The reference values and relevant information are provided in Tables 3.1 (ISO) and 3.3 (CEN). It should be noted that Table 3.1, taken from ISO 11228-1, also includes the reference mass of 23 kg as suggested, incidentally, in the original version of the RNLE.

In reality, the standards allow the user to decide which reference mass (or weight constant) to choose, depending on the population to be protected, the local legislation, and the user's own cultural approach. Once the reference mass has been decided, the methodology for using the RNLE remains unchanged, as featured in the standards.

Since this is an application manual, we will not go into complex issues such as the levels of protection afforded by different reference masses, or the wisdom of applying a single reference mass across the entire population rather than different masses for different population subsets.

Instead, this manual accommodates both approaches (provided the values are the ones envisaged by the standards, especially the ISO rules); therefore all the operating proposals contained in this volume may be applied based on a single weight constant (e.g., 23 kg, as suggested by the original RNLE) or on the weight constants (or reference masses) suggested by the standards as a function of different healthy working populations, as will be explained later.

3.2 ISO 11228-1: *ERGONOMICS—MANUAL HANDLING: PART 1: LIFTING AND CARRYING*

3.2.1 Introduction

The ISO 11228 family or series consists of three sections, under the general heading "Ergonomics—Manual Handling":

- Part 1: Lifting and Carrying
- Part 2: Pushing and Pulling
- Part 3: Handling of Low Loads at High Frequency

The three parts were prepared by Technical Committee ISO/TC 159, Ergonomics, Subcommittee SC 3, Anthropometry and Biomechanics. Part 1 of 11228 first appeared in 2003 [ISO, 2003], while Parts 2 and 3 were published in 2007 [ISO, 2007a, 2007b].

TABLE 3.1
Worksheet C.1: Reference Mass (m_{ref}) for Different Populations (ISO 11228-1)

Field of Application	m_{ref} (kg)	Percentage of User Population Protected: Female and Male	Female	Male	Population Group	
Nonoccupational use	5	Data not available			Children and the elderly	Total population
	10	99	99	99	General domestic population	
Professional use	15	95	90	99	General working population, including the young and old	General working population
	20					
	23					
	25	85	70	90	Adult working population	
	30	See note			Specialised working population	Specialised working population under special circumstances
	35					
	40					

Note: Special circumstances. While every effort should be made to avoid manual handling activities or reduce the risks to the lowest possible levels, there may be exceptional circumstances where the reference mass may exceed 25 kg (e.g., where technological developments or interventions are not sufficiently advanced). In these exceptional circumstances, increased attention and consideration must be given to the education and training of the individual (e.g., specialised knowledge concerning risk identification and risk reduction), the working conditions that prevail, and the capabilities of the individual.

ISO 11228-1 is the first international standard on the manual handling of loads. The purpose of ISO 11228-1 is to set recommended limits for the mass of objects being manually handled, taking into account factors such as working postures and the frequency and duration of lifting tasks, as well as the amount of effort that workers exert when carrying out activities associated with manual handling.

The standard applies to the manual handling of objects

> with a mass of at least 3 kilograms, at a moderate walking speed (from 0.5 m/s to 1.0 m/s).

The standard does not include holding objects, pushing or pulling objects, lifting with one hand, manual handling while seated, and lifting by two or more people (although some useful indications are provided).

The following documents are referenced: ISO/IEC Guide 51, *Safety Aspects—Guidelines for Their Inclusion in Standards*; ISO 7250:1996, *Basic Human Body Measurements for Technological Design*; ISO 14121, *Safety of Machinery—Principles of Risk Assessment*; and EN 1005-2, *Safety of Machinery—Human Physical Performance—Part 2: Manual Handling of Machinery and Component Parts of Machinery*. The referenced documents are indispensable for interpreting and applying this document.

The standard sets out the terms and definitions used in the document: *Manual handling*, *manual lifting*, and *manual lowering* are terms that refer to the handling of not only objects but also people or animals (although in reality, only general principles can be applied to the handling of live objects, rather than the risk computation tools supplied by the standard).

The *ideal posture* for manual handling is defined as:

> Standing symmetrically and upright, keeping the horizontal distance between the centre of mass of the object being handled and the centre of mass of the worker less than 0.25 m, and the height of the grip less than 0.25 m above knuckle height.

The *ideal conditions* for manual handling are defined as:

> conditions that include ideal posture for manual handling, a firm grip on the object in neutral wrist posture, and favorable environmental conditions.

Repetitive handling is defined as:

> handling an object more than once every 5 min.

3.2.2 Ergonomic Approach

ISO 11228-1 begins by providing information for evaluating work that involves manual lifting and carrying. If manual handling cannot be avoided, the hazards for a worker's health and safety must be assessed. The procedure proposed here provides a step-by-step approach, in which the evaluator judges the interrelated aspects of the various tasks.

As recommended by ISO 14121 and EN 1005-2 (which is illustrated below) the procedure suggests breaking the assessment down into four stages:

1. Hazard recognition
2. Risk identification
3. Risk estimation
4. Risk evaluation

3.2.3 Risk Evaluation

Figure 3.1 illustrates the procedure for examining the variables to be considered in manual lifting and carrying. The analysis and evaluation of manual lifting (and possibly also carrying tasks) encompass the five steps shown below.

3.2.3.1 Step 1: Nonrepetitive (Occasional) Manual Lifting under Ideal Conditions

Assuming conditions are ideal, an initial screening of occasional manual lifting (i.e., with a frequency of less than one lift every 5 min) requires the determination of the object's mass using Table 3.1 (as presented in Annex C, ISO 11228-1). In other words, for occasional lifts, the limits shown in Table 3.1 should not be exceeded, taking into account the characteristics of the reference population. Moreover, it should be noted that the values shown in the figure are the starting limits for the next steps (2 and 3).

3.2.3.2 Step 2: Repetitive Manual Handling under Ideal Conditions

Under ideal conditions, repetitive manual lifting tasks require the determination of the object mass as well as the lifting frequency. The ratio of mass to frequency is presented in Figure 3.2, with two possible scenarios: The first refers to short lifting durations of up to 1 h, and the second to moderate lifting durations of between 1 and 2 h. For durations of over 2 h, go to step 3.

The graph shows that the absolute maximal lifting frequency is 15 lifts per minute, for lifting durations of up to 1 h per day and object mass of up to 7 kg. Under ideal conditions, if the requirements relative to step 1 or 2 are satisfied, that completes the assessment and the risk is defined as acceptable. Otherwise, continue to step 3.

3.2.3.3 Step 3: Recommended Limits for Mass, Frequency, and Object Position

For this step, ISO 11228-1 recommends applying the RNLE equation (details in Annex A, paragraph A.7, ISO 11228-1); the relevant variables are explored in greater detail in subsequent chapters. After verifying certain assumptions, the calculation for the limit is valid under the following conditions:

- Smooth lifting with both hands, without jerking
- No tasks where the worker is partly supported (e.g., one foot not on the floor)
- Width of the object 0.75 m or less

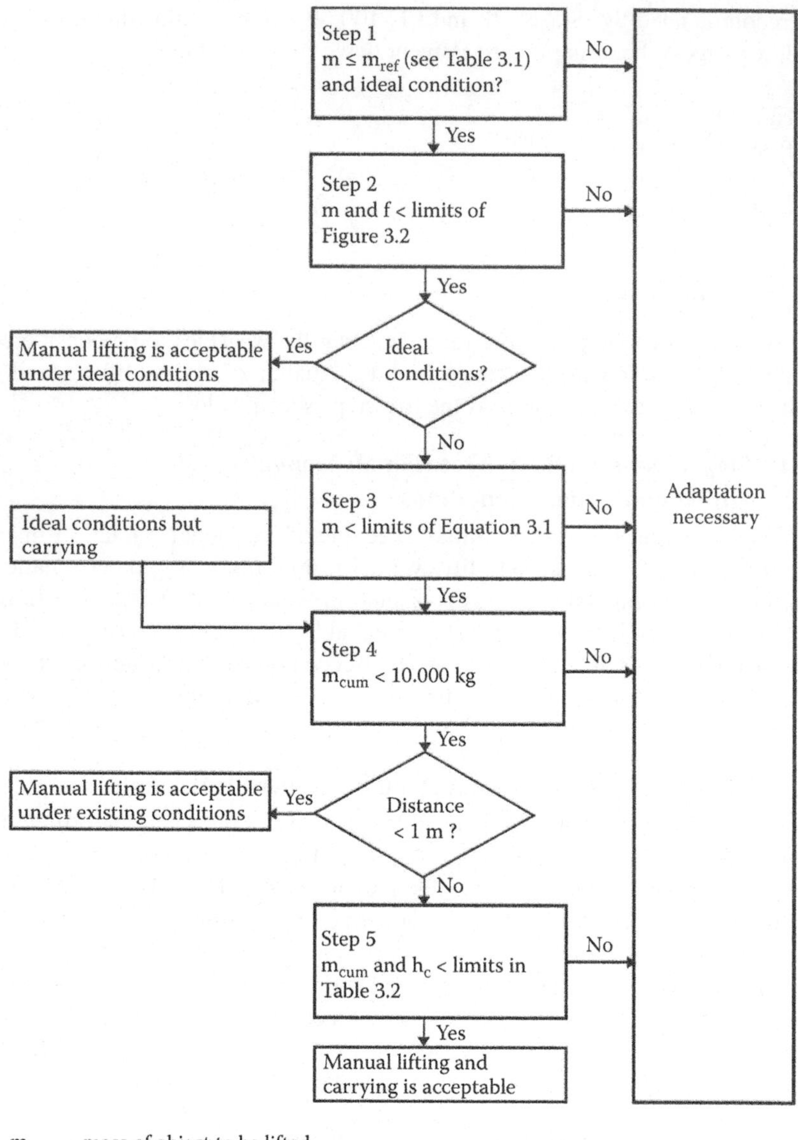

m mass of object to be lifted
m_{ref} reference mass for identified user population group
f frequency
m_{cum} cumulative mass
h_c distance (of carrying)

FIGURE 3.1 Step model (ISO 11228-1).

- Unrestricted lifting postures
- Good coupling, i.e., secure grip and low shoe-floor slip potential
- Favourable environmental conditions

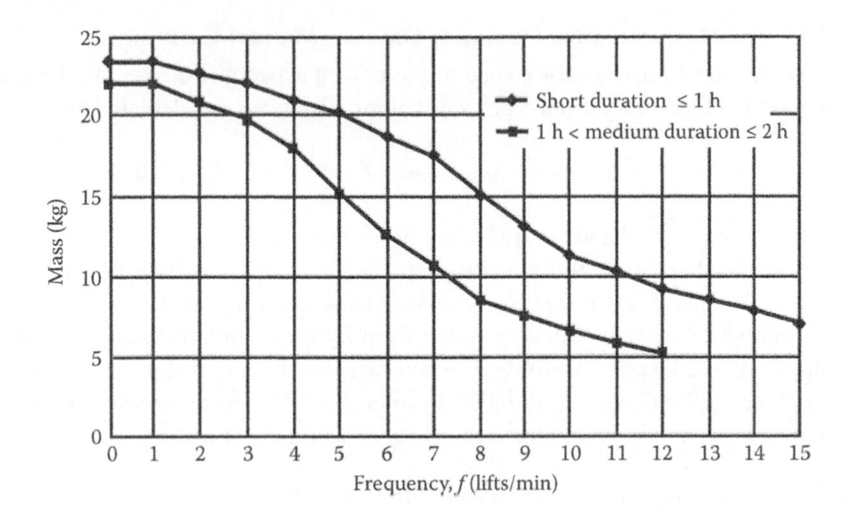

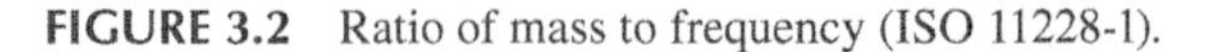

FIGURE 3.2 Ratio of mass to frequency (ISO 11228-1).

The limit for the mass of the object to be lifted is derived using the following equation:

$$m \leq m_{\text{ref}} \times h_M \times v_M \times d_M \times \alpha_M \times f_M \times c_M \quad (3.1)$$

where m_{ref} is the reference mass for the identified user population group (Table 3.1), h_M is the horizontal distance multiplier, v_M is the vertical location multiplier, d_M is the vertical displacement multiplier, α_M is the asymmetry multiplier, f_M is the frequency multiplier, and c_M is the coupling multiplier for the quality of gripping.

$$h_M = 0.25/h \text{ if } h \leq 0.25,\ h_M = 1 \text{ if } h > 0.63,\ h_M = 0 \quad (3.2)$$

$$v_M = 1 - 0.3 \times |0.75 - v| \text{ if } v < 0,\ v_M = 0.78 \text{ if } v > 1.75,\ v_M = 0 \quad (3.3)$$

$$d_M = 0.82 + 0.045/d \text{ if } d < 0.25,\ d_M = 1 \text{ if } d > 1.75,\ d_M = 0 \quad (3.4)$$

$$\alpha_M = 1 - (0.003\,2 \times \alpha) \text{ if } \alpha > 135°,\ \alpha_M = 0 \quad (3.5)$$

m_{ref} is obtained, as described above, from Table 3.1; the equations for obtaining h_M, v_M, d_M, and α_M are contained in the standard, and f_M and c_M are obtained from Tables A.1, A.2, and A.3 in Annex A from the Standard.

If the *recommended limit* for manual handling derived from the equation is exceeded (i.e., if the ratio of weight lifted to recommended weight is above 1), then the task must be redesigned, changing the mass or frequency of lifting or the duration of the task or position of the object. This therefore is the condition that defines the presence or acceptability of risk. ISO 11228-1 does not identify risk zones (green, yellow, red, as will be seen in EN 1005-2) but only a limit that, if exceeded, indicates the existence of risk.

3.2.3.4 Step 4: Cumulative Mass (per Day) for Manual Carrying

Step 4 of the procedure introduces one of the newest aspects of the standard, which is the concept of cumulative mass per day. Cumulative mass is calculated as

$$m_{\text{cum}} = m \times f \tag{3.6}$$

where m is mass carried and f is frequency of carrying.

These two values are both limited in steps 1 and 2: The mass should not exceed 25 kg, and the frequency of carrying should not exceed 15 times per minute. When the carrying distance is less than 10 m, the limit for the cumulative mass of manual carrying is 10,000 kg per 8 h. However, when carrying distance is longer (say, 20 m), this limit should be decreased to 6,000 kg (per 8 h). The corresponding limits for cumulative mass are also provided for periods of 1 min and 1 h.

3.2.3.5 Step 5: Recommended Limits for Cumulative Mass Related to Carrying Distance

Table 3.2 summarises the cumulative mass limits related to carrying distance and carrying frequency. Under unfavourable environmental conditions, or when lifting or lowering tasks take place closer to the ground, for instance, below knee level, or require lifting the arms above shoulder level, the recommended limits for cumulative mass for carrying in Table 3.2 should be reduced by at least one-third.

3.2.4 ISO 11228-1 Annex A

ISO 11228-1 Annex A indicates the ergonomic approach for removing or reducing the risk of manual handling injury. In the introduction, it is stated that ergonomics focuses on the design of work and its accommodation of human needs and physical and mental capabilities, considering manual handling tasks in their entirety and taking into account a range of relevant factors.

This approach is based on analysing manual handling tasks and evaluating the need for them. First, it is important to ask whether the manual handling of objects could be eliminated altogether. If not, the design of the work must consider the type of task, the layout of the handling area, and the organisation of the work.

The handling area should be adequate for avoiding the need for stretching the arms away from the body, twisting, bending, and flexing. The floor or ground should be level, and clear of obstacles. The best height for heavy objects is between mid-thigh and chest height, with lighter items being stored above or below this area.

The work should be organised so as to avoid awkward postures (see ISO 11226 for static working postures). Moreover, since the frequency of handling an object can influence the risk of back injury, consideration should be given to whether the handler has adequate opportunities for rest (i.e., momentary pauses or breaks from work) or recovery (i.e., changing to another task with no manual handling).

ISO 11228-1 also analyses situations in which *two or more people handle an object* whose weight is beyond the capability of one person. The weight of an object that can be safely handled by a team of two people is lower than the sum of what one

TABLE 3.2
Recommended Limits for Cumulative Mass Related to Carrying Distance, for the General Working Population (ISO 11228-1)

Carrying Distance	Carrying Frequency (f_{max})	Cumulative Mass (m_{max})			Examples
m	actions/min	kg/min	kg/h	kg/8 h	of Product ($f \times m$)
20	1	15	750	6,000	5 kg × 3 times/min
					15 kg × 1 time/min
					25 kg × 0.5 time/min
10	2	30	1,500	10,000	5 kg × 6 times/min
					15 kg × 2 times/min
					25 kg × 1 time/min
4	4	60	3,000	10,000	5 kg × 12 times/min
					15 kg × 4 times/min
					25 kg × 1 time/min
2	5	75	4,500	10,000	5 kg × 15 times/min
					15 kg × 5 times/min
					25 kg × 1 time/min
1	8	120	7,200	10,000	5 kg × 15 times/min
					15 kg × 8 times/min
					25 kg × 1 time/min

Note: In the calculation of the cumulative mass, a reference mass of 15 kg and a frequency of carrying of 15 times/min are used for the general working population.

The total cumulative mass of lifting and manual carrying should never exceed 10,000 kg/day, whichever is the daily duration of work.

23 kg is included in the 25 kg mass.

person could cope with individually. The standard defines the capability of a two-person team as two-thirds of the sum of their individual capabilities. For a three-person team, the capability is half the sum of their individual capabilities.

The *design of the object* must also be taken into consideration. The object to be handled may constitute a hazard if it does not have appropriate handgrips, or if its size or shape make it difficult to handle (e.g., if the width of the object exceeds the width of the worker's shoulders).

The *manual handling of people or animals* is dealt with in paragraph A.5 of the annex. Skill and experience are required to lift live objects, which present specific well-known problems.

As far as *environmental conditions* such as lighting, noise, and climate are concerned, safe handling requires levels to be within "tolerable levels." ISO 7730 is mentioned with specific reference to thermal comfort requirements.

At the end of Annex A there are a few remarks regarding the physical characteristics of people performing manual lifting tasks. The ability to perform manual lifting and carrying tasks will differ depending on *individual physical capabilities.*

Gender is also a factor: In general, the lifting strength of healthy women is up to two-thirds that of men. However, the range of strength and ability is large and will mean that some women can deal more safely with heavier objects than some men. However, there are situations in which manual lifting or carrying cannot be avoided, and may place particular demands on the physical capability of the worker, regardless of gender.

Age is another factor that must be taken into account. A younger individual may be stronger but less experienced, and therefore at risk of adopting hazardous lifting practices. Older individuals may be more susceptible to injury because parts of their musculoskeletal system have become less elastic. ISO 11228-1 states that the reduction in physical capability becomes more significant after the age of 45.

The section on individual characteristics ends with paragraph A.9, which states how crucial information and training are in reducing manual handling injuries.

ISO 11228-1 also includes Annexes B and C. Annex B presents two examples of manual handling of objects, analysed using the flowchart in Figure 3.1. Annex C includes Worksheet C.1, and gives reference masses for different populations (as described in step 1 and Table 3.1).

3.3 EN 1005-2: *SAFETY OF MACHINERY—HUMAN PHYSICAL PERFORMANCE PART 2: MANUAL HANDLING OF MACHINERY AND COMPONENT PARTS OF MACHINERY*

The EN 1005 series *Safety of Machinery—Human Physical Performance* is divided into five parts: EN 1005-2 is the second part.

EN 1005-1:2001. *Safety of Machinery—Human Physical Performance—Terms and Definitions* [CEN, 2001]

EN 1005-2:2003. *Safety of Machinery—Human Physical Performance—Part 2: Manual Handling of Machinery and Component Parts of Machinery* [CEN, 2003]

EN 1005-3:2002. *Safety of Machinery—Human Physical Performance—Recommended Force Limits for Machinery Operation* [CEN, 2002b]

EN 1005-4:2005. *Safety of Machinery—Human Physical Performance—Part 4: Evaluation of Working Postures and Movements in Relation to Machinery* [CEN, 2004]

EN 1005-5:2007. *Safety of Machinery—Human Physical Performance—Part 5: Risk Assessment for Repetitive Handling at High Frequency* [CEN, 2007]

The first four parts are considered harmonised standards for the design of machinery in compliance with the European directives for the safety of machinery (Machinery Directive 98/37/EC and, more recently, 2006/42/CE).

EN 1005-2 deals explicitly with the manual lifting and handling of loads, providing binding operating conditions to ensure machines' designs avoid handling risk or reduce the risks to the lowest possible level. Failure to comply (as for all harmonised standards) implies that the machine itself is noncompliant and therefore not qualified to carry the CE label.

This standard was first published in 2003 and the latest edition came out in 2009.

Normative references include EN 292-2 and EN 614-1 for basic concepts and ergonomic design principles, EN 1005-1 and EN 1070 for terms and definitions, and EN 1050 (now also ISO 14121) for the principles of risk assessment.

3.3.1 Approach to Risk Assessment

Paragraph 4.3 illustrates the rationale behind the model for assessing risk. The risk assessment model proposes three methods that share the same theoretical basis but feature different levels of complexity in their application.

- Method 1 is a quick screening method.
- Method 2 is slightly more complex but still simple enough to apply, and should be used if method 1 indicates risks.
- Method 3: If methods 1 and 2 do not suffice, method 3 must be applied. It is more thorough and is supplemented by additional risk factors not present in the other two methods. Clearly, the most efficient approach is to begin the risk assessment by applying method 1, then go on to methods 2 and 3 if the assumptions or operational situations identified in method 1 are not met.

Figure 3.3 of the standard presents the flowchart depicting the approach toward risk assessment.

The next paragraph in the standard (4.3.2) presents recommendations for hazard identification, risk estimation, risk evaluation, and risk reduction by correct machinery design. The indications refer to the mass, distribution, stability, and size of the objects, grips, and handles, picking and placing points of components and materials on machinery, frequency of operation, one-handed handling, or handling by two persons.

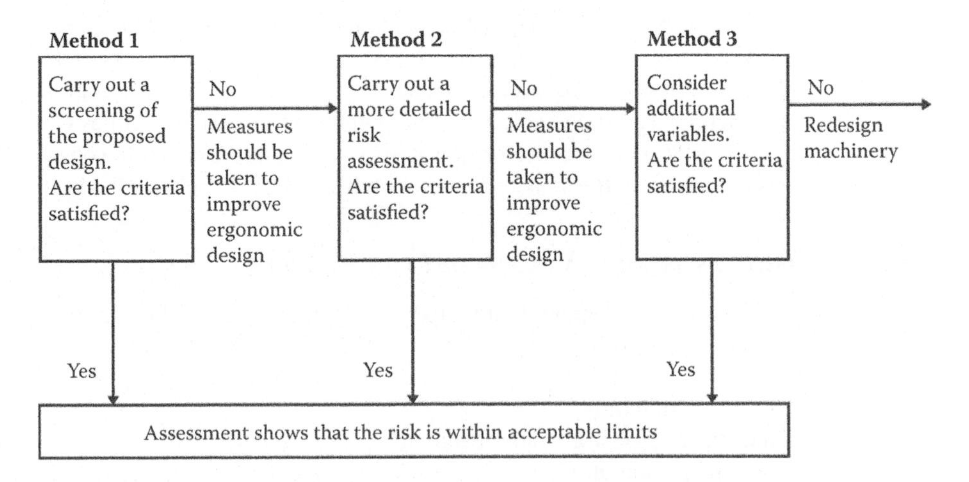

FIGURE 3.3 Flowchart identifying the approach to assessment in EN 1005-2.

TABLE 3.3
Reference Mass (m_{ref}) in Relation to the Intended User Population (EN 1005-2)

Field of Application	m_{ref} (kg)	Percentage of: Female and Male	Percentage of: Female	Percentage of: Male	Population Group	
Domestic use[a]	5	Data not available			Children and the elderly	Total population
	10	99	99	99	General domestic population	
Professional use (general)[b]	15	95	90	99	General working population, including the young and old	General working population
	25	85	70	90	Adult working population	
Professional use (exceptional)[c]	30 35 40	Data not available			Special working population	Special working population

[a] When designing a machine for domestic use, 10 kg should be used as a general reference mass in the risk assessment. If children and elderly people are included in the intended user population, the reference mass should be lowered to 5 kg.

[b] When designing a machine for professional use, a reference mass of 25 kg should not be exceeded in general.

[c] While every effort should be made to avoid manual handling activities or reduce the risks to the lowest possible level, there may be exceptional circumstances where the reference mass might exceed 25 kg (e.g., where technological developments or interventions are not sufficiently advanced). Under these special conditions other measures have to be taken to control the risk according to EN 614-1 (e.g., technical aids, instructions, or special training for the intended operator group).

The standard goes on to provide a risk assessment model comprised of three separate steps for each of the aforementioned methods:

Step 1 refers to determining the reference mass as a function of the population who will use the machine, based on Table 3.1 (from both the main text and Annex C).

Step 2 proposes a risk assessment performed at different levels, based on the operational situations contained in the standard, and applying three different methods.

3.3.2 Method 1: Screening by Means of Critical Values

To apply this quick screening method, first it has to be decided whether the handling operation is ideal or not, i.e., if the following criteria are met:

- Two-handed operation only
- Unrestricted standing posture and movements
- Handling by one person only

- Smooth lifting
- Good coupling between the hands and the objects handled
- Good coupling between the feet and the floor
- Manual handling activities, other than lifting, are minimal
- The objects to be lifted are not cold, hot, or contaminated
- Moderate ambient thermal environment

If all criteria are met, then one of the three following scenarios can be selected:

- *Critical mass*: The load handled does not exceed 70% of the reference mass, and the frequency is no more than 1 lift every 5 min.
- *Vertical displacement of the critical mass*: The load handled does not exceed 60% of the reference mass, and the frequency is no more than 1 lift every 5 min.
- *Critical frequency*: There are two alternatives: The load handled does not exceed 30% of the reference mass with a frequency of 5 lifts every minute, or the load handled does not exceed 50% of the reference mass with a frequency of 2.5 lifts every minute.

These criteria, together with a vertical displacement of the load equal to or less than 25 cm, between hip and shoulder height, with the trunk upright and not rotated and the load kept close to the body, describe acceptable operational situations, and the assessment may be regarded as completed.

Conversely, method 2 should be applied.

3.3.3 Method 2: Estimation by Tables (Worksheets)

After identifying the appropriate reference mass for the intended user population (Table 3.3), it has to be determined if the following criteria are met:

- Two-handed operation only
- Unrestricted standing posture and movements
- Handling by one person only
- Smooth lifting
- Good coupling between the feet and the floor
- Manual handling activities, other than lifting, are minimal
- The objects to be lifted are not cold, hot, or contaminated
- Moderate ambient thermal environment

If one or more of these criteria are not met, then method 3 should be applied; otherwise it is possible to assess the risk level by computing the recommended mass limit (*RML2*) and the risk index (*RI*) by comparing the actual mass with the recommended mass:

$$RI = \text{actual mass}/RML2 \tag{3.7}$$

TABLE 3.4
Worksheet Indicating Multipliers for Calculation of the Recommended Mass Limit (*RML*2) in EN 1005-2 (Method 2)

			Vertical Multiplier (VM)				
Vertical location (cm)	0	25	50	75	100	130	>175
Multiplier	0.78	0.85	0.93	1.00	0.93	0.84	0.00
			Distance Multiplier (DM)				
Vertical displacement (cm)	25	30	40	50	70	100	>175
Multiplier	1.00	0.97	0.93	0.91	0.88	0.87	0.00
			Horizontal Multiplier (HM)				
Horizontal location (cm)	25	30	40	50	56	60	>63
Multiplier	1.00	0.83	0.63	0.50	0.45	0.42	0.00
			Asymmetric Multiplier (AM)				
Angle of asymmetry (°)	0	30	60	90	120	135	>135
Multiplier	1.00	0.90	0.81	0.71	0.62	0.57	0.00
			Coupling Multiplier (CM)				
Quality of grip	Good			Fair			Poor
Description	See Table 4.2			See Table 4.2			See Table 4.2
Multiplier	1.00			0.95			0.90

Frequency Multiplier (FM)

	Frequency						
Hz	0.0033	0.0166	0.0666	0.1000	0.1500	0.2000	>0.2500
Lifts/min	0.20	1	4	6	9	12	> 15
Duration (d) $d \leq 1$ h	1.00	0.94	0.84	0.75	0.52	0.37	0.00
1 h $\leq d \leq$ 2 h	0.95	0.88	0.72	0.50	0.30	0.00	0.00
2 h $< d \leq$ 8 h	0.85	0.75	0.45	0.27	0.00	0.00	0.00

The recommended mass limit is calculated as follows:

$$RML2 = m_{ref} \times VM \times DM \times HM \times AM \times CM \times FM \quad (3.8)$$

For method 2, the multipliers are summarised in the worksheet shown in Table 3.4; this worksheet is very similar, and not by chance, to the one originally proposed by the EPM Research Unit in the Italian guidelines [Grieco et al., 1997].

3.3.4 Method 3: Calculation by Formula

If one or more of the criteria for using method 2 are not met or if the result suggests that risk is present or significant, method 3 should be applied. Again, the first step is to identify the reference mass using Table 3.3; then it needs to be decided if the following criteria are met:

- Unrestricted standing posture and movements
- Smooth lifting
- Good coupling between the feet and the floor
- The objects to be lifted are not cold, hot, or contaminated
- Moderate ambient thermal environment

If all the criteria are met, the risk assessment can be performed by calculating the recommended reference mass using the following equation:

$$RML = m_{ref} \times VM \times DM \times HM \times AM \times CM \times FM \times OM \times PM \times AT \quad (3.9)$$

In this case, the multipliers are not summarised in a worksheet, but should be obtained by applying the following formulas:

$$V_M = 1 - 0.003\ |V - 75|;\ \text{if } V < 0 \text{ cm},\ VM = 0.78;\ \text{if } V > 175 \text{ cm},\ VM = 0 \quad (3.10)$$

$$DM = 0.82 + 4.5/D;\ \text{if } D < 25 \text{ cm},\ D_M = 1;\ \text{if } D > 175 \text{ cm},\ DM = 0 \quad (3.11)$$

$$HM = 25/H;\ \text{if } H < 25 \text{ cm},\ HM = 1;\ \text{if } H > 63 \text{ cm},\ HM = 0 \quad (3.12)$$

$$AM = 1 - (0.003\ 2A);\ \text{if } A > 135°,\ AM = 0 \quad (3.13)$$

where m_{ref} is the reference mass from Table 3.3, in kg; V is the vertical location of the load, in cm; D is the vertical displacement of the load, in cm; H is the horizontal location of the load, in cm; A is the angle of asymmetry, in degrees; CM is the coupling multiplier from a specific table not provided here (see also Table 4.2, next chapter); and FM is the frequency multiplier from a specific table not provided here (see also Table 4.2, next chapter).

Method 3 introduces three additional multipliers (to original RNLE) that consider one-handed operation, two-person operation, and the presence of additional physically demanding tasks. Also in method 3, both the coupling multiplier and frequency multiplier are determined by two specific tables (see also Tables 4.2 and 4.4, next chapter).

OM is one-handed operation; in this case $OM = 0.6$, otherwise $OM = 1.0$.

PM is two-person operation; in this case $PM = 0.85$, otherwise $PM = 1.0$.

AT is the additional physically demanding task; in this case $AT = 0.8$, otherwise $AT = 1.0$.

3.3.4.1 Step 3

Based on the results of step 2, step 3 determines the necessary remedial actions. The risk index obtained by methods 2 and 3 should be compared with Table 3.5.

The standard ends with the required actions for risk reduction, which can be summarised as follows:

- No action if the risk is tolerable.
- If the risk is significant, the machinery must be redesigned, or the risk must be brought within acceptable limits by analysing the situation using a more complex method.
- If risk is present, then the machinery must be redesigned. The workplace can be redesigned/improved by modifying situations that lead to low multipliers.

TABLE 3.5
Risk Index and Its Interpretation (EN 1005-2)

Risk Index	Risk	Zone	Remedial Actions
$RI \leq 0.85$	Tolerable	Green	None.
$0.85 < RI < 1.0$	Significant	Yellow	It is recommended to redesign the machinery or ensure that the risk is tolerable.
$RI \geq 1.0$	Present	Red	Redesign is necessary. The design can be improved by changing the situations that lead to low multipliers.

TABLE 3.6
Population Percentages Protected in Relation to Measurement Criteria and Object Mass

Options	Psychophysical Data Indicating Tolerability Capacity	Measurements of Forces Indicating Limits	Measurements of the Maximum Metabolic Ability Limits
10 kg	99% (F + M)	99% (F + M)	99% (F + M)
	99% F	99% F	99% F
	99.9% M	99.9% M	99.9% M
20 kg	95% (F + M)	95% (F + M)	95% (F + M)
	90% F	90% F	80–85% F
	99.9% M	99.9% M	99% M
25 kg	85% (F + M)	85% (F + M)	85% (F + M)
	75% F	72–75% F	70% F
	99.9% M	99.9% M	99% M

Note: F = female, M = male.

If method 1 (step 3) detects the presence of risk, the designer is free to redesign immediately, without performing the risk assessment in method 2.

Three informative annexes are also provided.

Annex A contains an important recommendation: The worksheet in Table 3.6 indicates the results of studies to identify physical capability limits or maximum lifting tolerability in relation to the general working population, for three different types of studies and results: psychophysical data indicating tolerability capacity, physiological data with measurement of forces indicating limits, and measurements of the maximum metabolic ability limits. Representative sample populations were studied to obtain the reference masses and percentages of population protected (F = females, M = males).

Annex B contains indications regarding thermal comfort; here too, EN ISO 7730 should be applied. The recommended limits for climate and thermal conditions are 19–26°C for temperature, 30–70% for humidity, and ≤0.2 m/s for air velocity in closed environments.

Annex C contains the worksheet for reference masses seen previously (Table 3.3), in addition to three operational situations for risk assessment, including all the guidance for applying the three steps and the three methods proposed by the standard: critical values for method 1, operational worksheet for method 2, and calculations and worksheets for method 3.

3.4 OTHER STANDARDS IN THE ISO 11228 SERIES

3.4.1 ISO 11228-2: *Ergonomics—Manual Handling—Part 2: Pushing and Pulling*

Pain, fatigue, and disorders of the musculoskeletal system may be caused not only by lifting objects but also by pushing and pulling loads. ISO 11228, Part 2 indicates how to determine the potential hazards and risks associated with whole-body pushing and pulling.

The standard recommends two methods: The first is easier and is based primarily on the psychophysical tables designed by Snook and Ciriello [1991], which have been used for this purpose since 1991 as implemented by data from Mital, Nicholson, and Ayoub in their book [1993]. The second is considerably more complex and includes a procedure for determining whole-body pushing and pulling limits according to the characteristics of the specific population assigned to perform the task; in this case, the specific characteristics need to be identified and described.

The focus of this chapter is on the first method, which is easier to apply and more practical.

3.4.1.1 Introduction

The limits recommended by the standard apply to the healthy adult working population and provide reasonable protection to the majority of this population. The indications are based on three types of experimental studies: musculoskeletal loading, discomfort/pain, and endurance/fatigue.

Pushing and pulling activities are restricted to the following:

- Whole-body force exertions (i.e., while standing/walking)
- Actions performed by one person
- Forces applied by two hands
- Forces used to move or restrain an object
- Forces applied in a smooth and controlled way
- Forces applied without the use of external support(s) to objects located in front of the operator
- Forces applied in an upright position (not sitting)

For the purposes of the assessment, the following terms and definitions apply:

- *Pulling*: Human physical effort where the motive force is in front of the body and directed toward the body.
- *Pushing*: Human physical effort where the motive force is directed to the front of, and away from, the operator's body.

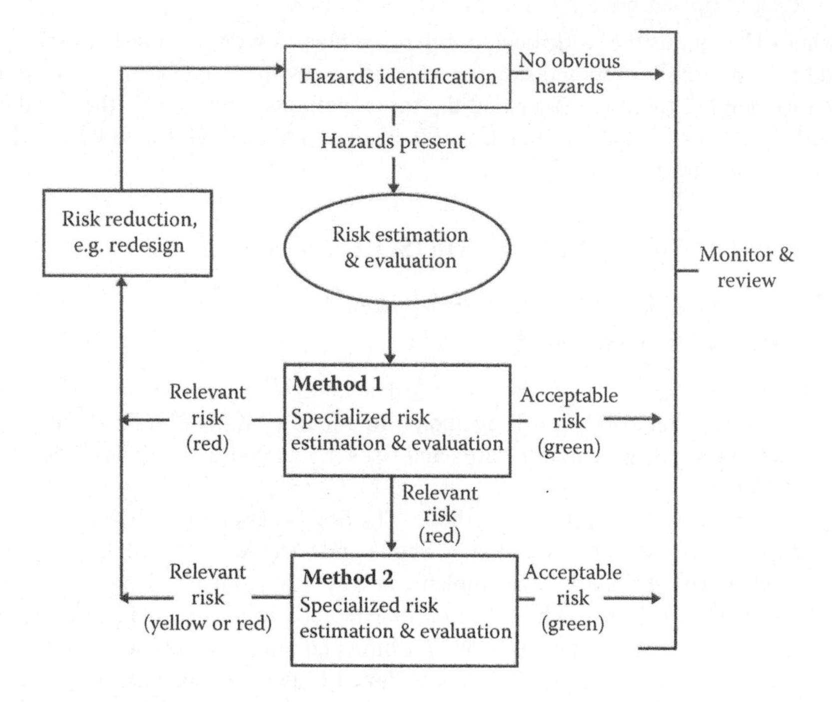

FIGURE 3.4 ISO 11228-2: Risk assessment model.

- *Initial force*: Force applied to set an object in motion.
- *Sustained force*: Force applied to keep an object in motion.

In the introduction, ISO 11228-2 includes several recommendations, such as that *hazardous* manual handling tasks should be avoided wherever possible. Appropriate job and workplace design can contribute toward achieving this goal.

Risk assessment is comprised of the following three steps:

- Hazard identification in pushing and pulling tasks
- Estimation of the consequent risk
- Evaluation of the actual risk

The risk assessment model proposed by the standard is shown in Figure 3.4.

3.4.1.2 Hazard Identification

The following eight interacting factors may combine to create hazards in tasks that involve pushing and pulling:

- Force
- Posture
- Frequency and duration
- Distance
- Characteristics of the object

- Environmental conditions
- Individual characteristics
- Work organisation

Initial and sustained forces are those required to start and maintain the movement of the object; obviously it would be preferable to avoid frequent starting, stopping, and manoeuvring of the object; similarly, jerky movements should be avoided. Regarding posture: awkward postures often lead to decreased abilities for force exertions. The operator should adopt a natural, balanced posture when applying initial or sustained push/pull forces, and minimise the forces acting on the back. Twisted, lateral bent and flexed trunk postures should be avoided. The hands that apply the force should be close to the body's centre of gravity, thus neither too high nor too low.

When pushing and pulling, both duration and frequency should be considered. High repetitive force exertions (i.e., continuous stopping and starting of pushes or pulls) will result in more frequent *initial forces* (rather than sustained force) and should be avoided.

Long distances may be more fatiguing to the operator. Specific mechanical aids should be used to manage long distances. The object to be pushed or pulled should also be suitably designed. If the object has wheels, they should be of a suitable diameter and material for the object and considering specific conditions of use (e.g., floor surfaces). Unfavourable environmental conditions include the presence of slopes or ramps and wet, slippery, or irregular surfaces. Hazards may be enhanced by vibration, inappropriate lighting, and hot or cold environments.

Individual skills and capabilities, level of training, age, gender, and health status may increase or decrease the risk of injury. Work schedules that include other load handling and lifting activities, in addition to pushing and pulling, may contribute to biomechanical overload during the course of the workday and should be defined based on all the tasks performed.

3.4.1.3 Risk Estimation and Risk Assessment

As stated above, ISO 11228-2 envisages two different evaluation methods:

Method 1: This provides a simple task analysis using psychophysical tables with which to identify the reference values for acceptable initial and sustained forces, as a function of the factors that constitute the task, such as handle height, distance moved, and frequency of push/pull tasks for males and females. Method 1 not only recommends reference values, but also sets out the steps for reducing the risk of injury.

Annex A includes checklists to be used for recording information about the task:

- *Checklist 1:* General screening for presence/absence of risk. If risk is detected, the second checklist should be used.
- *Checklist 2:* Identification of working posture personnel involved and operating flows.

- *Checklist 3:* This includes the form for evaluating potential risk factors. Essentially, this table is a checklist with a number of questions about the six elements analysed prior to measuring force, i.e., task to be performed, load or object to be moved, characteristics of any wheels or castors, work environment, individual operator capabilities, and any other factors present. Completing the checklist provides a yes/no answer, where yes indicates the presence of risk. If risk is detected, the reasons for defining the hazard must be entered, together with suggestions and possible remedial actions.
- *Checklist 4:* This checklist is for determining initial and sustained forces. To make an accurate evaluation, the following must be determined:
  1. Handle height
  2. Distance pushed or pulled
  3. Frequency of push/pull actions, both initial and sustained
  4. Worker population, i.e., composition: all male (use male limits) or all female (use female limits) or mixed male/female (use female limits)
- *Determining acceptable forces.* Then, using the forms provided in the standard, it is necessary to determine the acceptable initial and sustained force, in order to protect 90% of the worker population broken down into males, females, or mixed males/females.
- *Measuring initial and sustained force.* Assessments should also measure initial and sustained forces. To perform these measurements correctly, clear practical instructions are included in Annex D.
- *Rating risk.* Once all the input data have been collected and the various forces have been measured, they can be compared with the risk rating provided.

Method 1 describes only two conditions: risk or no risk (red/green zone). However, there are three possible conditions that also involve the results reported in checklist 3, for the existence of possible other risk factors.

1. If the force measured is greater than the recommended maximum force, then risk is present and the task is classified as red.
2. If the force measured is lower than the recommended maximum force, but a predominant number of risk factors is identified from the checklist, then risk is present and the task is still classified as red.
3. Otherwise, the risk is classified as green.

Following remedial actions, if the overall assessment is red or the risk level is difficult to evaluate, then risk reduction actions must be implemented, or for a more in-depth analysis, method 2 should be followed.

Method 2: The second method proposed by ISO 11228-2 is complex. It consists of a procedure for analytically determining whole-body pushing and pulling force limits, according to specific characteristics of the population and the task. These values are obtained from an in-depth study.

Method 2 is divided into four parts:

Part A—Determination of muscle force limits: This part determines force limits based on static strength measurements and adjusts those forces according to population characteristics (i.e., age, gender, and stature) and the requirements of the task (i.e., frequency, duration, and distance of push/pull task).

Part B—Determination of skeletal force limits: This part takes into account push/pull tasks resulting in high lumbar spinal compressive forces and adjusts push/pull forces according to spinal compression limits for age and gender.

Part C—Determination of maximum forces permitted: This part identifies the maximum force applicable using muscle force limits.

Part D—Determination of safety limits: This part defines the safety limits by determining the risk multiplier *mr*. Unlike method 1, the risk multiplier *mr* is divided into three risk zones (green, yellow, and red) as follows:

1. *Green zone* (acceptable risk), $mr < 0.85$: The risk of disease or injury is negligible or is at an acceptably low level for the entire operator population. No action is required.
2. *Yellow zone* (conditionally acceptable risk), $0.85 < m_r \leq 1.0$: There is a risk of disease or injury that cannot be neglected for the entire operator population or part of it. The risk shall be further estimated, analysed together with contributory risk factors, and followed as soon as possible by redesign. Where redesign is not possible, other measures to control the risk shall be taken.
3. *Red zone* (risk not acceptable), $1.0 < m_r$: There is a considerable risk of disease or injury that cannot be neglected for the operator population. Immediate action to reduce the risk (e.g., redesign, work organisation, worker instruction and training) is necessary.

Method 2 is complex and its use is recommended only in special cases, such as particular worker populations whose characteristics differ considerably from the standard working population samples, or when an extremely in-depth analysis is required for medicolegal reasons.

There are 6 annexes to ISO 11228-2, identified by letters A to F.

As mentioned, the first two annexes describe the application of methods 1 and 2. The annexes include all the reference tables and formulas.

Annex C is informative and proposes a specific approach toward risk reduction. The variables and concepts presented in this annex are also referred to in paragraph 3.2.1 of the standard "Hazard Identification." For each variable (repetitive movement, task, posture, workplace, work organisation, design of objects or tools handled, design of the work environment, individual worker capabilities), the standard proposes improvements or suggestions for reducing risk factors.

Annex D suggests an approach for measuring pushing and pulling forces.

Annex E presents practical examples for applying methods 1 and 2.

Annex F defines the method for determining combined strength distribution for a particular reference group.

3.4.2 ISO 11228-3: Ergonomics—Manual Handling—Part 3: Handling of Low Loads at High Frequency

At the end of this chapter, readers will find the third and last standard in the ISO 11228 family, which closely resembles EN 1005, Part 5.

The contents of these two standards are very similar, and for this reason reference will be made primarily to ISO 11128-3.

The two standards deal with a special form of manual load handling, which is the handling of low loads at high frequency; if performed continuously, such tasks may cause pain and fatigue, which could lead to musculoskeletal disorders, also called repetitive strain injuries.

Risk factors for these tasks, defined as repetitive work, include frequency of actions, exposure duration, awkward postures and movements of body segments, hand forces associated with the task, work organisation, inadequate recovery times, and level of training/skill. Additional factors may include environmental factors such as climate and vibrations, and the way certain tasks are performed (continuous striking actions or the use of hands for balancing).

This part of the standard also derives from in-depth experimental studies. It provides guidance for identifying and evaluating risk factors related to the handling of low loads at high frequency (repetitive movements and force exerted by the upper limbs). The recommendations apply to the adult working population in general, and are intended to give reasonable protection for nearly all healthy adults. The standard should be used as a reference guide for all those involved in the design or redesign of work, jobs, and products.

3.4.2.1 Risk Evaluation

The procedure for evaluating risk follows the approach given by ISO 14121, and consists of four separate steps for both assessing risk and reducing risk:

1. Hazard identification
2. Risk estimation
3. Risk assessment
4. Risk reduction

3.4.2.2 Step 1: Hazard Identification

The first step is to identify whether hazards exist that may expose individuals to risk. The following hazards should be considered:

- Repetition (frequently performed movements)
- Posture and movement
- Force
- Duration and insufficient recovery
- Object characteristics
- Vibration and impact forces
- Environmental condition
- Work organisation
- Psychosocial factors
- Individuals

For each factor there are recommendations for identifying their existence during working activities. If any of these factors are detected, the next step to be performed is the risk estimation.

3.4.2.3 Step 2: Method 1—Risk Estimation

Risk estimation is performed by a simple risk assessment of jobs, broken down into four parts: (1) preliminary information, (2) hazard identification, (3) overall evaluation of the risk, and (4) remedial action to be taken. To perform these four steps, Annex B proposes the methods and checklists for collecting information in order to identify hazards. There may be three types of results, rated as green, yellow, or red. The three risk zones are defined as:

1. *Green zone (acceptable risk)*: The risk of disease or injury is negligible or is at an acceptably low level for the entire working population. No action is required.
2. *Yellow zone (conditionally acceptable risk)*: There is a risk of disease or injury that cannot be neglected for the entire working population or part of it. The risk shall be further estimated (using the more detailed assessment of method 2), analysed together with the contributory risk factors and followed as soon as possible by redesign. Where redesign is not possible, other measures to control the risk shall be taken.
3. *Red zone (not acceptable)*: There is a considerable risk of disease or injury that cannot be neglected for the working population. Immediate action to reduce the risk is necessary (e.g., redesign, work organisation, worker instruction and training).

Annex A to the standard offers more methods for evaluating risk due to repetitive movements and exertions, including the OCRA checklist, the Plibel, the OSHA checklist, and the QEC. The standard generally considers these methods as simple (and often empirical) screening methods, and some of them (notably the Ovako Working Posture Assessment System [OWAS] or Rapid Upper Limb Assessment [RULA]) are described as lending themselves to the study of awkward postures, but as unsuitable for evaluating risk caused by repetitive movements.

3.4.2.4 Step 3: Method 2—Detailed Risk Assessment

If method 1 reveals a risk estimate ranked as yellow or red, or if the job is comprised of more than one repetitive task, then a more detailed risk evaluation is required.

The preferred approach for performing a detailed risk assessment is the occupational repetitive action (OCRA) method. This method is recommended for more specific investigations because, based on current knowledge at the time of publishing the standard, it is viewed as the most comprehensive. In fact, the OCRA method analyses all the risk factors related to and associated with repetitive movements, and it is also applicable to multitask jobs, providing criteria (based on extensive epidemiological data) for predicting the onset of upper limb work-related musculoskeletal disorders (UL-WMSDs) in exposed populations.

The OCRA index is the ratio of the number of actual technical actions (ATAs) performed during a shift to the number of reference technical actions (RTAs), for each upper limb, specifically determined by the cycle being analysed.

TABLE 3.7
Ranges That Define Risk Levels

OCRA Index Value	Risk Level	Consequences
≤2.2	No risk	Acceptable: no consequences
2.3–3.5	Very low risk	Improve structural risk factors (posture, force, technical actions, etc.) or take other organisational measures
>3.5	Significant risk	Redesign tasks and workplaces according to priorities

OCRA = Number of actual technical actions performed in a shift (ATA)/Number of reference technical actions performed in a shift (RTA)

For the procedure defined by the OCRA index, see www.epmresearch.org.

The value of the resulting index is then compared with the ranges defining the risk level (Table 3.7).

Besides the OCRA index, which the standard describes as preferable to the others, mention is also made of other detailed risk evaluation methods for single task manual work, such as the Strain Index and the HAL/ACGIH TLV. In the standard, Annex D briefly describes these last two methods.

Annex A of Standard 11228-3 offers a few reasons for preferring the OCRA index over the other methods, while Annex C is entirely devoted to providing detailed explanations of the procedures used in the OCRA method.

3.5 AN APPLICATION DOCUMENT (ISO TR 12295) FOR THE STANDARDS IN THE ISO 11228 SERIES

After defining a series of technical norms for physical ergonomics with respect to working postures (ISO, 2000), manual lifting (ISO, 2003, 2007a), and repetitive manual tasks (ISO 2007b), the ISO is now preparing an application document (ISO TR 12295) to help explain the methods and procedures for implementing the methods set forth in the standards, and to help users apply them to real-life situations.

This TR (technical report) should be published by the end of 2012.

Several excerpts from the TR are provided here, especially with regard to the application of ISO 11228-1, which is of particular relevance for manual lifting tasks. These excerpts will help in understanding the recommended procedures for evaluating the lifting tasks presented in later chapters.

In general, it should be noted that the ISO application document (TR 12295) is intended to assist the user to decide which standards should be applied depending upon whether specific risks are present. Users will be required to answer a short series of practical "key enters" to assist them in applying the appropriate standard(s). The application document will also provide other information relevant for the practical application of methods and procedures presented or recommended in ISO 11228-1/2/3 with a special focus to situations where multiple manual tasks are performed by the same (group of) worker(s).

In general the application document will contain the following:

- Detailed definition of field of application of different standards
- Key enters (simple parametric hazard identification) to open different standards
- Systems for a quick assessment of exposure
- Updates of classification systems in Part 1 (lifting)
- Updates of the main selected methods used in the standards with particular reference to multitask analysis of lifting tasks
- Reference to websites relevant for applying the standards

3.5.1 Hazard Identification by Simple Key Enters

The key enters are designed to assess if there is any relevance of the basic conditions of the job to the specific standards (step 1). If there is relevance, then the user is directed to the applicable quick assessment questions (step 2) that will then give an approximated indication of the magnitude of the potential hazard, either low/no risk or the presence of critical (high-risk) conditions. If hazards are indicated, then the user is directed to the appropriate standard.

The key enters addressing the different parts of the standard are reported in Table 3.8.

TABLE 3.8
Procedure for Applying Standards: Key Questions (Step 1), from ISO TR 12295

1	**Application of ISO 11228-1**		
	Is there manual lifting or carrying of an object of 3 kg or more present?	No	Yes
	If no, this standard is not relevant, go to the next key question regarding the other standards.		
	If yes, go to step 2, quick assessment (Tables 3.9, 3.10, and 3.11).		
2	**Application of ISO 11228-2**		
	Is there manual whole-body pushing and pulling of loads present?	No	Yes
	If no, this standard is not relevant, go to the next key question regarding the other standards.		
	If yes, go to step 2, quick assessment (Tables 3.9, 3.10, and 3.11).		
3	**Application of ISO 11228-3**		
	Are there one or more repetitive tasks* of the upper limbs with a total duration of 1 h or more per shift?	No	Yes
	*where the definition of *repetitive task* is "task characterised by cycles or task during which the same working gestures are repeated for more than 50% of the time"		
	If no, this standard is not relevant, go to the other key question regarding the other standards.		
	If yes, go to step 2, quick assessment (Tables 3.9, 3.10, and 3.11).		
4	**Application of ISO 11226**		
	Are there working postures of the head/neck, trunk, or upper and lower limbs maintained for more than 4 s consecutively and repeated for a significant part of the working time?	No	Yes
	If no, this standard is not relevant.		
	If yes, go to step 2, quick assessment (Tables 3.9, 3.10, and 3.11).		

3.5.2 Quick Assessment (Step 2)

This step serves to make a rapid assessment of any potential risks, by answering a set of simple qualitative and quantitative questions.

Its purpose is simply detecting three possible conditions (outputs):

1. *Acceptable (green code):* No actions required.
2. *Critical (critical or purple code*): The work or process needs urgent redesigning.
3. *More detailed analysis is required*: A detailed estimate or assessment is necessary using the risk estimation and evaluation checklists described extensively by the standards.

As well, this step is often explicitly indicated in the aforesaid international technical standards, especially if used to quickly assess acceptable conditions. For example, in manual lifting tasks, steps 1 and 2 of ISO 11228-1 (or similarly, method 1 of EN 1005-2) provide a rapid method for assessing the acceptability of the object actually lifted and the relevant lifting frequency, in relation to the reference values supplied by the standards. On the other hand, to make a quick assessment of definitely critical conditions, it is possible to use the definitions and criteria included in the methods recommended by the standards, which indicate one or more risky elements: These may be objects weighing more than the recommended maximum limit, extreme load lifting areas, or very high lifting frequencies.

Regarding manual lifting and carrying tasks, a preliminary check of some adverse environmental, object, and organisational conditions is highly recommended since those conditions could represent an additional risk in manual handling and could cause inapplicability of step 3 in ISO 11228-1. This procedure is also recommended in EN 1005-2 and in European Directive 269/90. The preliminary check of acceptable conditions is based on Table 3.9.

To perform a quick assessment of acceptable conditions (low risk/green area) in manual lifting and carrying tasks, Table 3.10 should be used (it summarises steps 1 and 2 of ISO 11228-1): If any of the conditions are not met, refer directly to ISO 11228-1, step 3 (apply RNLE).

A quick assessment could also be used for identifying critical (or purple code) conditions (for lifting and carrying): If only one of the conditions is met, a critical situation in lifting or carrying is present, and an urgent ergonomics intervention is necessary.

These critical conditions, as reported in Table 3.11, are all clearly indicated in the original RNLE method and consequently in ISO 11228-1. The term *critical* means that the manual lifting of objects is practically not allowed.

As mentioned, the purpose of the quick assessment is to simply identify whether working conditions are acceptable or critical when there is a potential risk of biomechanical musculoskeletal overload. When either one of these conditions is met, a more detailed risk assessment is not necessary. However, in most cases neither condition emerges clearly; therefore it is necessary to carry out a risk assessment using the methods laid down by the reference standards (in this case, ISO 11228-1).

TABLE 3.9
Preliminary Check of Unfavourable Conditions (Preliminary Aspects) for Lifting and Carrying

Hazard	**Is the working environment unfavourable for manual lifting and carrying?**		
	Presence of extreme (low or high) temperature	No	Yes
	Presence of slippery, uneven, unstable floor	No	Yes
	Presence of insufficient space for lifting and carrying	No	Yes
Hazard	**Are there unfavourable object characteristics for manual lifting and carrying?**		
	The size of the object reduces the operator's view and hinders movement	No	Yes
	The load centre of gravity is not stable (e.g., liquids, items moving around inside of object)	No	Yes
	The object shape/configuration presents sharp edges, surfaces, or protrusions	No	Yes
	The contact surfaces are too cold or too hot	No	Yes
Hazard	**Does the task(s) with manual lifting or carrying last more than 8 h a day?**	No	Yes

If all the hazard conditions are no, continue the quick assessment.
If one is yes, apply ISO 11228-1.
The consequent specific additional risks have to be carefully considered to minimise these risks.

TABLE 3.10
Quick Assessment for Acceptable Conditions—Lifting and Carrying

Lifting: Quick Assessment: Acceptable Condition (Green Code)

3–5 kg	Asymmetry (e.g., body rotation, trunk twisting) is absent	No	Yes
	Load is maintained close to the body	No	Yes
	Load vertical displacement is between hips and shoulders	No	Yes
	Maximum permissible frequency: less than 5 lifts/min	No	Yes
5.1–10 kg	Asymmetry (e.g., body rotation, trunk twisting) is absent	No	Yes
	Load is maintained close to the body	No	Yes
	Load vertical displacement is between hips and shoulder	No	Yes
	Maximum permissible frequency: less than 1 lift/min	No	Yes
More than 10 kg	Loads more than 10 kg are not present	No	Yes

Carrying: Quick Assessment: Acceptable Condition (Green Code)

Consider the reported recommended cumulative mass (total kg carried in the considered period for the given distance): Is the effectively carried accumulative mass less than recommended values considering distances (more/less than 10 m) and periods (1 min, 1 h, 8 h)?

	For Less Than 10 m	For More Than 10 m		
8 h	10,000	6,000	No	Yes
1 h	1,500	750	No	Yes
1 min	30	15	No	Yes
	Awkward postures are not present		No	Yes

If all the listed conditions are yes, the examined task is in the green area (acceptable) and it is not necessary to continue the risk evaluation.
If one is no, apply ISO 11228-1.

TABLE 3.11
Quick Assessment for Critical Conditions—Lifting and Carrying

If only one of the following conditions is present, risk has to be considered as high and it is necessary to proceed with task redesign.

Critical Condition: Presence of Layout and Frequency Conditions Exceeding the Maximum Suggested		
Vertical location	Hands at the beginning/end of the manual lifting, higher than 175 cm or lower than 0 cm	Yes
Vertical displacement	The vertical distance between the origin and the destination of the lifted object is more than 175 cm	Yes
Horizontal distance	The horizontal distance between the lifted object and the body centre of gravity (medium point between the ankles) is more than 63 cm	Yes
Asymmetry	Asymmetry angle (upper body rotation) more than 135°	Yes
Frequency	More than 15 lifts/min in short duration (manual handling lasting no more than 60 min consecutively in the shift, followed by at least 60 min of break or light task)	Yes
	More than 12 lifts/min in medium duration (manual handling lasting no more than 120 min consecutively in the shift, followed by at least 30 min of break or light task)	Yes
	More than 8 lift/min in long duration (manual handling lasting more than 120 min consecutively in the shift)	Yes
Critical Condition: Presence of Loads Exceeding the Following Limits		
Males (18–45 years)	25 kg	Yes
Females (18–45 years)	20 kg	Yes
Males (<18 or >45 years)	20 kg	Yes
Females (<18 or >45 years)	15 kg	Yes
Critical Condition for Carrying: Presence of Cumulative Carried Mass Greater Than Those Indicated		
6,000 kg	Carrying distance 20 m or more in 8 h	Yes
10,000 kg	Carrying distance less than 20 m in 8 h	Yes

If only one answer is yes, a critical situation is present.
Proceed with assessment with ISO 11228-1 for identifying urgent corrective actions.

3.5.3 Annex A (Informative): Application Information for 11228-1

With this annex, TR 12295 provides additional application information regarding ISO 11228-1 and the recommended methods, also taking into account the latest advances reported in the literature.

At present, this information might not be directly included in the TR, and there has been a proposal to transfer it to a new version of ISO 11228-1.

Despite these limitations, a few details are presented here concerning the information contained in the current draft of the TR, since it could be useful for the purposes of this manual.

TABLE 3.12
Reference Masses (m_{ref}) Suggested in Annex A, ISO TR 12295

Working Population by Gender and Age	Reference Mass (m_{ref})
Men (18–45 years old)	25 kg
Women (18–45 years old)	20 kg
Men (<18 or >45 years old)	20 kg
Women (<18 or >45 years old)	15 kg

Note: 23 kg is the reference mass used in the U.S. National Institute of Occupational Safety and Health (NIOSH) lift equation, which is the source of the lifting analysis method used in ISO 112281. The use of 23 kg as the reference mass accommodates at least 99% of male healthy workers and at least 75% of female healthy workers at $LI = 1.0$.

3.5.3.1 Reference Mass

Table 3.1 in this manual (Worksheet C.1 in the standard) presents reference masses with estimated corresponding percentages of various user populations, who are protected when these reference masses are used in the lifting task assessment. Based on Table 3.1 and similar tables in other relevant standards (i.e., EN 1005-2), the reference mass reported in Table 3.12 could be adopted in relation to the gender/age of the working population.

3.5.3.2 Use of Lifting Index (LI)

Once the *recommended limit for mass* (mR) has been computed, starting from *reference mass* (m_{ref}) in Table 3.12 and using the procedures and equations given in ISO 11228-1, step 3, it is possible to compare the *actually lifted mass* (mA) with the resulting recommended limit (mR).

This comparison is made by checking if:

$$mA \leq mR \text{ (acceptable condition)}$$

$$mA > mR \text{ (not recommended condition)}$$

Alternatively, another widely used method is to compute the *lifting index* (LI). The lifting index is equal to the ratio between the *actually lifted mass* (mA) and the corresponding *recommended limit* (mR):

$$LI = mA/mR$$

With the lifting index, the classification of the results (green/red) according to ISO 11228-1, step 3, becomes:

If lifting index ≤ 1, acceptable condition
If lifting index > 1, not recommended condition

TABLE 3.13
Detailed Interpretation of LI Values as Suggested in ISO TR 12295

Lifting Index Value	Risk Level	Interpretation	Consequences
$LI \leq 1$	Acceptable	Exposure is acceptable for most members of the reference working population. Lifting conditions accommodate >90% of males and females, including younger and older.	Acceptable: No consequences
$1.1 < LI < 2.0$	Present; low level	A part of the adult industrial working population could be exposed to a low risk level.	Redesign tasks and workplaces according to priorities
$2.1 < LI < 3.0$	Present; significant level	A larger part of the adult industrial working population could be exposed to a significant risk level.	Redesign tasks and workplaces as soon as possible
$LI > 3.0$	Unacceptable	Absolutely unsuitable for most working populations. Only for exceptional circumstances, e.g., where technological developments or interventions are not sufficiently advanced. Under such exceptional circumstances, increased attention and consideration must be given to the education and training of the individual (e.g., specialised knowledge concerning risk identification and risk reduction).	Redesign tasks and workplaces immediately

For a helpful interpretation of the lifting index, especially when greater than 1 (bad condition), and to address the intervention priorities also with reference to the relevant scientific literature, reference can be made to Table 3.13.

3.5.3.3 Guidance on Multitask Lifting Analysis

This section acknowledges that multitask lifting analysis is often used for jobs performed in different ways than the single task jobs analysed with the traditional RNLE (reported in ISO 11228-1, step 3). In the literature, several methods have been proposed for analysing complex multitask lifting jobs, all based on the RNLE: composite (CLI) [Waters et al., 1994], sequential (SLI) [Waters et al., 2007], and more recently, variable lifting tasks (VLI) [Waters et al., 2009], [Colombini et al., 2009]. On this last type of task, the guidance devotes considerable space to proposals described in greater detail elsewhere in this manual, which may help users find useful documents and software programs online.

Therefore, this manual should be regarded as an application tool not only for ISO 11228-1 but also for the relevant "application document" reported in ISO TR 12295.

4 Procedures, Models, and Criteria for Evaluating Risk in Manual Lifting Jobs

4.1 THE ORGANISATIONAL ANALYSIS: DEFINING MANUAL LIFTING JOBS

To evaluate risk in manual lifting jobs it is first necessary to identify tasks and subtasks that call for the following actions:

1. Lifting and lowering objects (loads) with a mass of 3 kg or more (3 kg value was selected according to ISO 11228-1 [ISO, 2003] and EN 1005-2 [CEN, 2003])
2. Carrying loads possibly related to the lifting and lowering task
3. Pulling and pushing loads

By defining jobs and tasks that require manual lifting, lowering, carrying, pushing or pulling loads, it is possible to define which tasks require subsequent risk evaluation.

Therefore, before starting any risk assessment, it is necessary to conduct a careful organisational analysis to study the content of each shift. Several definitions may be helpful for this purpose:

1. Organised work or job: A set of organised working activities (tasks) performed during a work shift. It may be comprised of one or more tasks.
2. Manual load lifting task: Lifting and lowering objects weighing 3 kg or more.
3. Load carrying tasks: When an object remains lifted and is moved horizontally (1 m or more) by human force; this activity is generally related to lifting and lowering.
4. Manual load pulling or pushing tasks: Activity related to starting or stopping the movement of an object that either has or does not have wheels.
5. Tasks that do not involve the manual handling of loads or light work: These include all jobs that are sedentary, involve visual monitoring or light assembly jobs that do not require lifting or carrying weights of over 3 kg, or pulling/pushing loads.
6. Official or unofficial breaks: Uninterrupted periods spent away from working activities.

With regard to work shifts, the following must be described:

1. Total duration of the shift
2. Type, duration, and distribution of manual load handling jobs, broken down as indicated above
3. Type, duration, and distribution of other jobs that do not involve manual load handling
4. Type, duration, and distribution of breaks

In the study of manual load lifting activities, four types of tasks and relevant turnover can be operationally identified, with the following definitions and features:

- *Single task*: This task involves lifting objects of the same type (and the same weight) with no changes in the parameters (same position of the layout and body geometry) from one lift to the other both at the origin of the lifting and at the destination. In this case, the classic lifting index (LI) [Waters et al., 1993] method can be used.
- *Composite task*: Lifting objects that are generally of the same type (and weight) but over different geometries (for instance, grasping and moving objects from or to shelves at different heights or different horizontal distances). Each individual geometry (i.e., each combination of vertical height and horizontal distance at origin or destination) is called a subtask. In this case, the composite lifting index (CLI) [Water et al., 1994] can be computed following the specific procedure. However, it has been postulated that no more than 10–12 subtasks can be computed in this procedure, hence the need to introduce standardised simplifications (see Chapter 7).
- *Variable task*: This refers to lifting or lowering objects of different weights and over different geometries (vertical heights, horizontal distances) within the same time period. Different weight categories could be identified in this case. The handling of each separate weight category over each individual geometry is called a subtask. In this case the variable lifting index (VLI) is the calculation methodology to be used (see Chapter 8).
- *Sequential task*: When a daily shift includes several different lifting tasks (single, composite, or variable), each performed continuously for at least 30 min. Workers rotate between a series of single or composite or variable lifting rotation slots during a work shift. In this case the sequential lifting index (SLI) [Waters et al., 2007] is the calculation methodology to be used.

Figure 4.1 proposes a model for gathering organisational data referred to a work shift. It is a type of daily diary in which all the events of the day are described and timed in succession.

In computing the time of the manual load handling task, carrying times should initially also be included: The total time (lifting + carrying) is required to study the lifting frequency. Pushing and pulling times are indicated in separate boxes.

Work Shift Schedule																
Other tasks or breaks	MANUAL LIFTING TASK	Other tasks or breaks	Pushing/pulling task	Other tasks or breaks	MANUAL LIFTING TASK	Other tasks or breaks	Pushing/pulling task	Other tasks or breaks	MANUAL LIFTING TASK	Other tasks or breaks	Pushing/pulling task	Other tasks or breaks	MANUAL LIFTING TASK	Other tasks or breaks	Pushing/pulling task	Other tasks or breaks
	60		15		**120**	30	20		**60**		15	40	**120**			

FIGURE 4.1 A model for gathering work organisation data arranged across a work shift. The duration of manual lifting also includes carrying. The description lists all light tasks or breaks as well as pulling and pushing tasks in succession.

Time spent in light tasks, which do not involve manually lifting objects weighing over 3 kg or pushing/pulling tasks should be indicated together with breaks: Altogether, these represent recovery times for the lower back, as indicated by the RNLE method for evaluating lifting indexes.

This first model for describing organisational data is part of a worksheet for collecting the data needed to evaluate lifting indexes, designed as a guide for entering the data collected during inspections. The entire worksheet will be referred to from time to time throughout the various operational steps that appear in this book.

4.2 EVALUATING MANUAL LIFTING TASKS: GENERAL ASPECTS

The evaluation methods indicated by both ISO 11228-1 and EN 1005-2 largely correspond to the revised NIOSH lifting equation (RNLE) [Waters et al., 1993].

The revised NIOSH lifting equation, published in 1993 and updating a previous version of the approach that appeared in 1981, envisages a formula that integrates organisational, weight, and geometry (layout and body position) factors; its aim is to define the recommended weight limit (RWL) for a lifting activity. This recommended weight is then compared with the weight actually lifted to compute the lifting index (LI):

$$LI = L/RWL \tag{4.1}$$

where L = load weight and RWL = recommended weight limit.

The recommended weight limit (RWL) is estimated to start from the *maximum weight nearly all healthy workers should be able to lift under optimal conditions (load constant or reference mass)* reduced by the intervention of other relevant risks factors (multipliers or reduction factors).

4.3 THE CONCEPT OF REFERENCE MASSES

In the introduction of Chapter 3 a brief discussion was presented regarding the different approaches from NIOSH, ISO, and CEN when referring to the concept of reference masses to be used when applying the RNLE.

Substantially, while NIOSH suggests a single reference mass (23 kg or 51 lb) for the whole healthy adult working population (male and female), ISO and CEN standards suggest the use of different reference masses in relation to specific subgroups (considering gender and age) of the adult working population. Moreover, it is worth noting that in the United States most typically the user may chose a LI limit for the appropriate population as opposed to picking a reference mass (that is quite the same). This happens also when using Liberty Mutual psychophysical tables [Snook and Ciriello, 1991], which are proposed considering, at first step, gender aspects.

Once again we underline that this manual, in a pragmatic way, accommodates both approaches; therefore all the operating proposals contained in this volume may be applied based on a single load constant (e.g., 23 kg, as suggested by the original RNLE) or on different constants (or reference masses) suggested by the standards as a function of different components of the healthy working populations, as will be explained hereafter.

The load constant (reference masses) values can be derived from ISO 11228-1 and EN 1005-2. The relevant tables were presented in Chapter 3 (Tables 3.1 and 3.3), where the values, such as the reference masses for different population groups, appear to show differences that could cause confusion in choosing the reference weights, to be used as ideal masses for computing the recommended weight and thus the lifting index.

It is worth noting that the general aim of the regulations on ergonomics is to protect the physical and mental well-being and health of at least 90% of the reference population. This is the basis for choosing reference masses. Specifically, ISO 11228-1 states that using a reference mass of 25 kg, 95% of the healthy male adult working population should be protected.

Regarding the reference mass for other working populations (male and female, young and old), the aforementioned ISO standard is complicated, as it provides other reference values, such as 23, 20, and 15 kg for similar target populations and similar protection levels. The 23 kg value was chosen since it is the one primarily used in the United States (by the revised NIOSH lifting equation).

With regard to other parts of the working population, it may also be useful to refer to EN 1005-2: This regulation indicates tolerance limits (in its Annex A) for individual working populations. Specifically, Annex A in EN 1005-2 states that using a reference mass of 20 kg protects over 90% of the male working population (including both younger and older workers). Accordingly, the use of a reference mass of 20 kg for a population of both younger and older male workers is justified.

For the female working population, using 20 kg as the reference mass protects 85–90% of the adult population, while the protections levels are lower for younger and older women: Consequently, the proposal is to take 20 kg as the reference mass only for the adult female working population and 15 kg for the younger and older female population.

Based on the forgoing considerations, apart from using the reference mass of 23 kg for all the working population (as proposed in the original NIOSH approach), the reference masses for different subgroups (by gender and age) of the working populations could be, based on ISO 11228-1, those summarised in Table 4.1. The age cutoff of 45 years was selected since there is a specific indication in ISO 11228-1, but there are other indications in literature that it could be moved to 50 years.

TABLE 4.1
Reference Masses Based on ISO 11228-1 and EN 1005-2

Gender	Age Range	Reference Mass
Male	18–45	25 kg
Female	18–45	20 kg
Male	<18 or >45	20 kg
Female	<18 or >45	15 kg

The reference mass is the heaviest weight that can be lifted under ideal conditions (see later) and still accommodate almost 90% of the relevant population. Exceeding these weights during lifting, even only occasionally, stays for a critical condition: If that mass is exceeded, then something less than 90% of the relevant population will be protected.

Moreover, ISO 11228-1 and EN 1005-2 standards do not consider these reference masses as insurmountable limits, but rather as indicators for a substantial protection among the various populations.

In fact, masses of even over 25 kg are listed for populations defined as special: In certain jobs, such conditions may occur quite frequently (e.g., routine or special maintenance operations, etc.); therefore different measures must be adopted to keep risk levels under control (for instance, with the use of suitable devices, procedures, and training, health monitoring/surveillance, etc.).

In any case, the concept remains that exceeding these threshold values is an important indicator of risk situations. In this respect, the key questions described in Table 3.8 can be seen to be a useful tool for identifying the most significant risk factors, in the first step toward assessing risk.

4.4 THE EVALUATION METHOD AND FACTORS CONSIDERED

As mentioned, reference masses may be regarded as the maximum weight that can be lifted under ideal conditions. Lifting conditions (whether ideal or otherwise) are determined both by workplace layout and body positions (geometries) and by the way work is organised.

Specifically, the method proposed by the NIOSH defines the following multiplication factors:

- *Vertical multiplier* (VM): Considering the distance of the hands from the floor measured at the start and the end of lifting.
- *Distance multiplier* (DM): Considering the maximum vertical distance moved between the start and the end of lifting.
- *Horizontal multiplier* (HM): Considering the horizontal distance between the load and the body at the start and the end of lifting.
- *Asymmetric multiplier* (AM): Considering the angular measure of displacement of the load from the sagittal plane of the subject.

- *Coupling multiplier* (CM): Considers the quality of the hand-to-object coupling (the way loads are gripped).
- *Frequency multiplier* (FM): Considers the average number of lifts per minute relative to the work duration.

The mathematical product (Figure 4.2) of the reference masses selected and these multipliers produce the recommended weight limit (RWL) under actual lifting conditions. In the present context, ideal conditions are synthetically defined as those where all the multipliers have a value of 1, thus confirming, as RWL, the selected reference mass.

LC	**Load Constant or Reference Mass**		Maximum recommended weight under ideal lifting conditions
VM	**Vertical Multiplier**	 A	Vertical distance of the hands from the floor at the start/end of lifting
DM	**Distance Multiplier**	 B	Vertical distance of the load between the beginning and the end of lifting
HM	**Horizontal Multiplier**	 H	Horizontal distance between the load and the body at the start/end of lifting
AM	**Asymmetric Multiplier**	 Y	Angular measure of displacement of the load from the sagittal plane at the start/end of lifting
CM	**Coupling Multiplier**		Assessment of grip of the object
FM	**Frequency Multiplier**		Frequency of lifts per minute and duration

FIGURE 4.2 Multipliers used in the RNLE method for computing the recommended weight limit (RWL).

More generally, *ideal conditions* are defined by ISO 11228-1 as "conditions that include ideal posture for manual handling, a firm grip on the object in neutral wrist posture, and favourable environmental conditions."

The *ideal posture for manual handling* is defined as "standing symmetrically and upright, keeping the horizontal distance between the centre of mass of the object being handled and the centre of mass of the worker less than 0.25 m, and the height of the grip less than 0.25 m above knuckle height."

At last *repetitive handling* is defined as "handling an object more than once every 5 min"; consequently, ideal conditions are those with lifting frequencies equal to or less than 0.2/min for a consecutive duration equal to or less than 1 h.

It is interesting to note that the technical regulations, i.e., both EN 1005-2 and ISO 11228-1, identify different approach levels, from the simplest to the most complex, based on the aforementioned multipliers.

The first level is a fast method for verifying the presence of generally optimal conditions. This first approach serves first of all to verify the existence of ideal lifting geometries:

1. Erect upper body
2. Object moved within the space between the height of the hips and that of the shoulders
3. Load kept close to the body and not bulky in shape
4. Good grip of the object

Generally speaking, once it has been ascertained that these conditions are ideal, the existence of a relationship between the weight of the load and the frequency of lifting can be defined. If the lifting is not carried out under ideal conditions, or if the ratio of the frequency of lifting to the load actually lifted is not respected, the regulations call for the application of more detailed evaluation levels.

For more practical operational details about the first-level analyses, reference can be made to Section 3.5 on the use of key questions and quick evaluations.

Regarding the more detailed procedures to evaluate the risk, EN 1005-2 bases method 2 on a risk evaluation using predefined geometric and organisational lifting multipliers, while method 3 directly offers a precise calculation of the individual multipliers.

Standard ISO 11228-1 refers directly to the precise calculation of lifting multipliers (substantially according to original RNLE) and therefore coincides, albeit with some differences that will be explained later, with method 3 of EN 1005-2.

To simplify the use of the multipliers so as to determine the recommended weight limit, and to standardise the language used by the two regulations, from here on reference will be made to:

- *Method 2* for defining the simplified multiplier values as indicated in EN 1005-2.
- *Method 3* for defining the precise calculation methods for the same multipliers, as laid down by both regulations. In proposing multipliers, both methods are identical to the original RNLE.

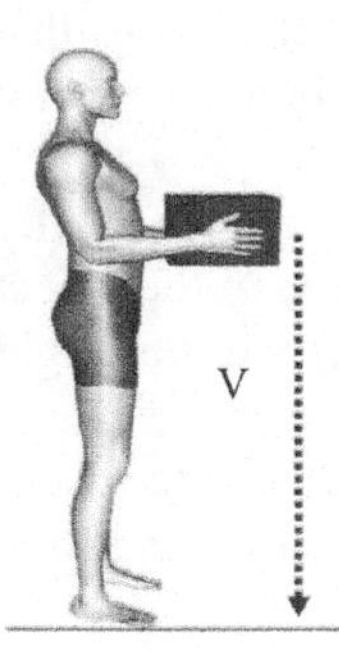

Vertical Multiplier (VM)
distance of the hands from the floor at the start (end) of lifting

The height of the hands above the floor (V) is measured vertically from the floor surface to the midpoint between the hand grasps.
The extremes of this height are given by the floor surface and the upper limit for vertical reach for lifting (175 cm).
The optimum level for this multiplier (VM = 1) is 75 cm. (knuckle height).
The value of VM decreases as the distance from this optimum level either increases or decreases.

Application limits
If the height is more than 175 cm, VM = 0.
If the height is less than 0 cm, VM = 0.

Vertical multipliers (VM) – simplified procedure:

Height V(cm)	0	25	50	75	100	125	150	>175 or < 0
VM	**0.77**	**0.85**	**0.93**	**1**	**0.93**	**0.85**	**0.75**	**0.00**

The formula for computing VM:
VM =1 – (0.003 × |V – 75|)

Where V = height of hands from floor in cm.

FIGURE 4.3 Vertical multiplier (VM).

4.4.1 Vertical Multiplier (VM)

This is defined as the distance from the midpoint between both hands and the floor (at the start or end of the lifting) (Figure 4.3).

The distance of the hands above the floor (V—vertical location) is measured vertically from the floor to the midpoint between the hand grasps. The extremes of this distance are given by the floor surface and the upper limit for vertical reach for lifting (i.e., 175 cm).

The optimum level for this multiplier (VM = 1.00) is obtained when the vertical height is 75 cm (the knuckle height is anatomical). The value of VM decreases as the distance from this optimum level either increases or decreases. If the height is over 175 cm or lower than the floor (i.e., <0 cm), VM becomes critical.

In method 2 (multipliers per areas), VM can be directly determined from the values shown in Figure 4.3. However, it is always possible to interpolate the multiplier for intermediate heights with respect to those listed in the table.

For both regulations, method 3—unlike the previous method—determines VM through the application of the original NIOSH formula:

$$VM = 1 - (0.003\ [V - 75]) \tag{4.2}$$

where V is distance of the hands from the floor in cm. In this case, rather than obtaining the multiplier from a predefined table, it is computed directly.

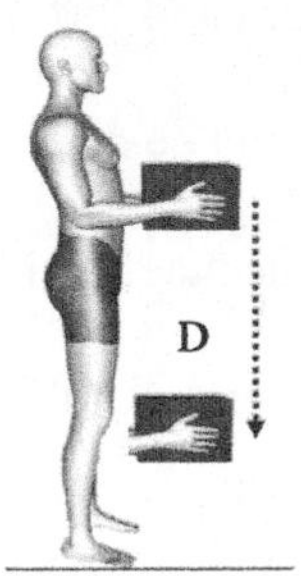

Distance Multiplier (DM)
difference in height of the hands between lifting and lowering

The vertical travel distance (D) is given by the vertical shift of the hands during lifting. This distance can be measured as the difference between the height of the hands at the origin and the destination of the lift.

If the object being lifted has to be passed over an obstacle, the vertical travel distance will be given by the difference between the highest point reached to pass over the obstacle and the height of the hands at the start (or end) of lifting and/or lowering.

The optimal distance is maximum 25 cm (DM = 1).

Application limits

If the travel distance is more than 175 cm, VM = 0.

Distance multipliers (DM) – simplified schedule:

Distance D(cm)	25	30	40	50	70	100	170	>175
DM	1	**0.97**	**0.93**	**0.91**	**0.88**	**0.87**	**0.86**	**0.00**

The formula for computing DM:

DM = 0.82 + (4.5/D)

Where D = vertical distance in cm.

FIGURE 4.4 Distance multiplier (DM).

4.4.2 Distance Multiplier (DM)

The vertical travel distance (D) is defined as the vertical travel distance of the hands between the origin and destination of the lift (Figure 4.4).

The vertical travel distance is given by the vertical shift of the hands during lifting. This distance can be measured as an absolute value equal to the difference between the height of the hands at the origin and the destination of the lift.

If the object being lifted has to be passed over an obstacle, the vertical travel distance will be given by the difference between the highest position of the hands and the lower position, i.e., the highest (worst case) of the following three calculation methods (taking absolute values):

1. Difference in hand height between the destination and the origin of the lift
2. Difference between the height of the obstacle and the height of the hands at the origin of the lift
3. Difference between the height of the obstacle and the height of the hands at the destination of the lift

The optimum level for this multiplier ($DM = 1.00$) is obtained when the vertical travel distance is less than or equal to 25 cm. DM decreases as the vertical travel distance increases (in cm).

It is not possible to obtain vertical travel distances greater than 175 cm due to the distance restrictions imposed on the vertical multiplier VM ($VM \geq 0$ cm and $VM \leq$ 175 cm) indicated in the previous paragraph.

In method 2 (simplified multipliers) the vertical travel distance DM can be determined directly from the values shown in Figure 4.4.

However, it is always possible to interpolate the multiplier for intermediate vertical travel distances with respect to those listed in the table.

Method 3, which is valid for both ISO 11228-1 and EN 1005-2, determines the vertical travel distance by applying the following formula:

$$DM = 0.82 + (4.5/D) \tag{4.3}$$

where D is vertical travel distance in cm.

4.4.3 Horizontal Multiplier (HM)

The horizontal multiplier is defined as the horizontal location (H) from the (vertical) projection of the midpoint of the hand grasp (i.e., load centre) to the body centre (Figure 4.5).

The horizontal location (H) is measured from the midpoint of the line joining the inner ankle bones (body centre) to a point projected on the floor directly below the midpoint of the hand grasps (load centre). When the load centre does not coincide with the hand grasp (as when lifting a long container where the centre of mass of the

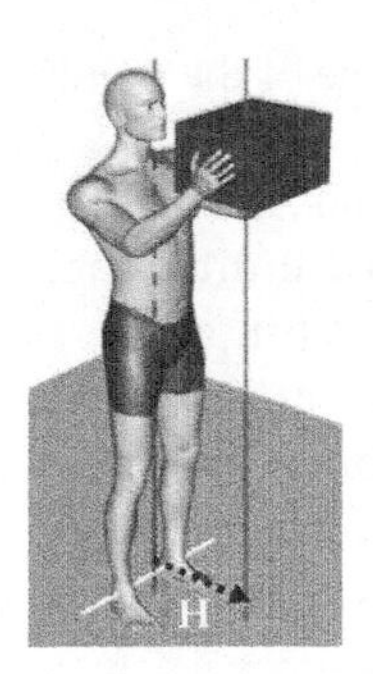

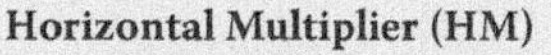

Horizontal Multiplier (HM)
horizontal distance between the load and the body

The horizontal location (H) is measured from the midpoint of the line joining the inner ankle bones (body center) to a point projected on the floor; it is the horizontal distance between the body center and the load center.

If the load center does not lie at the center of an imaginary line midpoint between the hand grasp, then measure the true distance from the center of the load and not the hand grasp.

The optimal distance is equal to or less than 25 cm (HM = 1).

Application limits
If the distance is more than 63 cm., VM = 0.

Horizontal multipliers (HM) – simplified schedule:

Distance cm	25	30	40	50	55	60	>63
HM	**1**	**0.83**	**0.63**	**0.50**	**0.45**	**0.42**	**0.00**

The formula for computing HM:
HM = 25/H

Where H = horizontal distance in cm.

FIGURE 4.5 Horizontal multiplier (HM).

object lies distant from the area of the hand grasp), the distance from the body centre is used as the true distance from the load centre.

The optimum level for this multiplier ($HM = 1.00$) is obtained when the horizontal location is less than or equal to 25 cm. HM decreases as the horizontal location increases.

If the horizontal location is over 63 cm, the HM assumes a critical value of 0. In method 2, HM can be determined directly from the values shown in Figure 4.5. However, it is always possible to interpolate the multiplier for intermediate horizontal distances with respect to those listed in the table.

Method 3 determines the horizontal location by applying the following formula:

$$HM = 25/H \qquad (4.4)$$

where H is the horizontal location in cm.

4.4.4 Asymmetric Multiplier (AM)

The angle of asymmetry is defined by the location of the load relative to the worker's mid-sagittal plane (Figure 4.6). Using references to the position of the feet or the worker's body twist may be erroneous.

The sagittal line is defined as the line passing through the mid-sagittal plane, dividing the body into two equal halves when the worker is in the neutral position, i.e., standing with no body twist.

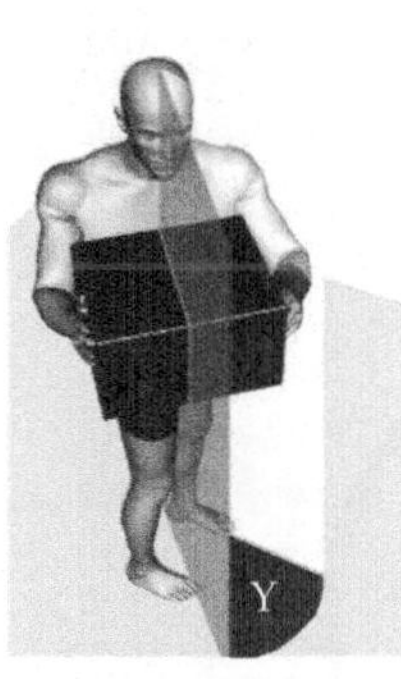

Asymmetric Multiplier (AM)
angle of body twist

The angle of asymmetry (Y) is the angle formed by the asymmetry line and the mid-sagittal line. The asymmetry line ideally joins the midpoint between the ankles and the point projected on the floor below the midpoint of the hand grasps at the start (or end) of lifting.

The angle of asymmetry is not defined by the position of the feet or the twisting of the worker's body twist, but by the location of the load relative to the worker's mid-sagittal plane.

The optimal condition is equal to or less than 25 degrees (AM = 1).

Application limits
If the legs or torso are twisted more than 135 degrees, VM = 0.

Asymmetry multipliers (AM) – simplified schedule:

Angle°	0	30	60	90	120	135	>135
AM	**1**	**0.90**	**0.81**	**0.71**	**0.62**	**0.57**	**0.00**

The formula for computing AM:
AM = 1 – (0.0032 Y) Where Y = angle of asymmetry in degrees

FIGURE 4.6 Asymmetry multiplier (AM).

The asymmetric angle ranges from 0° ($AM = 1$) to 135° ($AM = 0.57$). For angles of over 135° the AM assumes a critical value of 0.

The observation and measurement of the asymmetry angle requires a few additional explanations.

In general, asymmetric lifting may be necessary if the origin and destination of the lift are at an angle to each other. In this case, the asymmetry may be intrinsic to the task or determined by the worker's individual behaviour. However, it is important to eliminate the causes of the asymmetry, particularly in the risk reduction phase.

In calculating the asymmetry angle, there are four general situations that may require different evaluation methods:

1. *Worker at workstation with total freedom of movement*: The origin and destination may be at an angle to each other, but stepping is possible. This situation is often observed when lift frequency is rather low and the distance between the origin and destination of the lift allows for stepping.
2. *Worker at workstation with partial freedom of movement* (Figure 4.7): The origin and destination are at an angle of 180° to each other; the worker twists the upper body at various angles. To overcome the difficulty of measuring the various twist angles observed, a constant may be used that is equal to half of the total angle (90° in the example). This situation is often observed when lift frequency is relatively high and the distance between the origin and destination of the lift does not allow for stepping. The asymmetry multiplier in this case is $AM = 0.71$.
3. *Worker at workstation with no freedom to move but in a symmetrical position* (Figure 4.8): The origin and destination points are at an angle to each other and the worker's starting position is midway between the start and

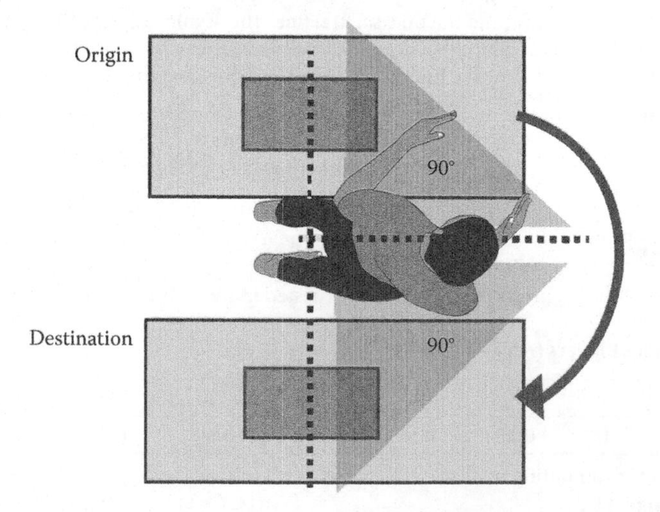

FIGURE 4.7 Angle of asymmetry with partial freedom of movement (origin and destination at an angle of 180°).

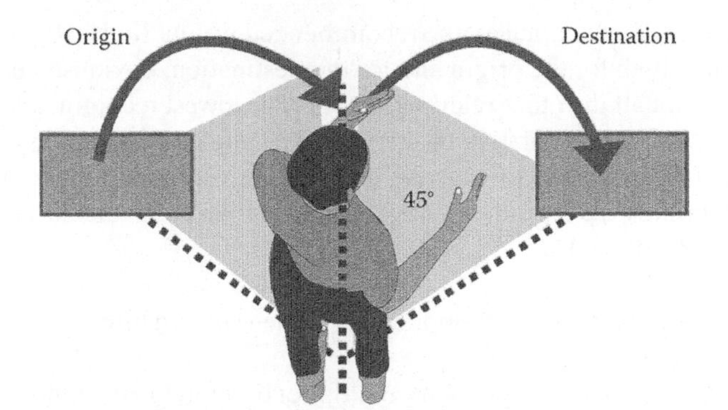

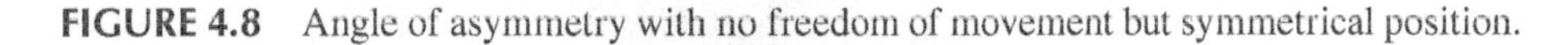

FIGURE 4.8 Angle of asymmetry with no freedom of movement but symmetrical position.

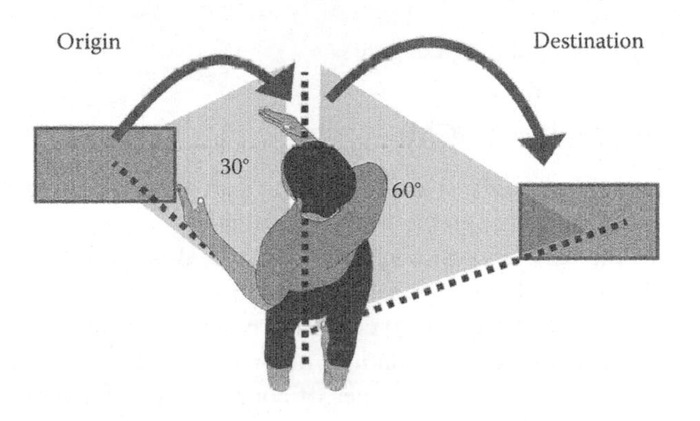

FIGURE 4.9 Angle of asymmetry with no freedom of movement and nonsymmetrical position.

end of the lift. In this case the angles between the plane of asymmetry at the origin/sagittal plane and sagittal plane/plane of asymmetry at destination coincide and represent the asymmetry angle. The task begins from the midpoint between the start and end of the lift. For example, if origin and destination are set at an angle of 90°, the worker turns 45° toward the origin to grasp the object, turns toward the destination point, passing through the starting point (0°), and then reaches the destination point with a 45° turn. The angle of asymmetry at the origin coincides with the angle of asymmetry at the destination, and will be 45°. The asymmetry multiplier in this case is $AM = 0.86$.

4. *Worker at workstation with no freedom to move and an asymmetric position* (Figure 4.9): The origin and destination are at an angle to each other and the worker adopts an asymmetric starting position midway between the start and end of the lift with respect to the sagittal plane. In the example, the worker turns 30° toward the origin to grasp the object, then turns toward the destination point, passing through the starting point (0°), and reaches the destination point with a 60° turn. Since the angles between the asymmetry line at the origin/sagittal plane and the sagittal plane/plane of asymmetry at

destination do not coincide, the recommended weight limit (RWL) must be calculated both for the origin and for the destination, obviously also taking into account all the other relevant factors. The lowest recommended weight will be set as representative of the task. The angle of asymmetry at origin will therefore be 30°, giving rise to an asymmetry multiplier: $AM_{\text{origin}} = 0.90$. Similarly the asymmetry angle at destination is 60° and its asymmetry multiplier will be $AM_{\text{destination}} = 0.81$.

In method 2 the asymmetry multiplier can be determined directly from the values shown in Figure 4.6.

Method 3 determines the asymmetry multiplier by applying the following formula:

$$AM = 1 - (0.0032y) \tag{4.5}$$

where y is the asymmetry angle in degrees.

4.4.5 Coupling Multiplier (CM)

The coupling multiplier is defined as the way the load is coupled or gripped, and is based on an assessment of the quality of the coupling.

The quality of the hand-to-load coupling is classified, as shown in Table 4.2, as good ($CM = 1$), fair ($CM = 0.95$), or poor ($CM = 0.9$).

To more correctly assess the type of coupling, based on the aforesaid definitions, reference must be made to recommendations in the literature on ergonomics relating to anthropometric measurements and types of couplings as a function of the object to be handled.

TABLE 4.2
Coupling Multiplier (CM)

Grasp Quality	Good	Fair	Poor
Description	Load length ≤ 40 cm; load height ≤ 30 cm; suitable handles or handhold cutouts. Object and handles have a simple configuration that allows the hand to easily grip it and wrap around it without significant wrist deviations.	Load length ≤ 40 cm; load height ≤ 30 cm; unsuitable handles or handhold cutouts or grip requiring the hand to be flexed 90°. Object with a simple configuration that allows the hand to grip it with a flexion of 90° but without significant wrist deviations.	Load length > 40 cm or load height > 30 cm, or objects difficult to handle or soft or asymmetric centre of mass or unstable contents or difficult to grip or requiring the use of gloves.
CM	1.00	0.95	0.90

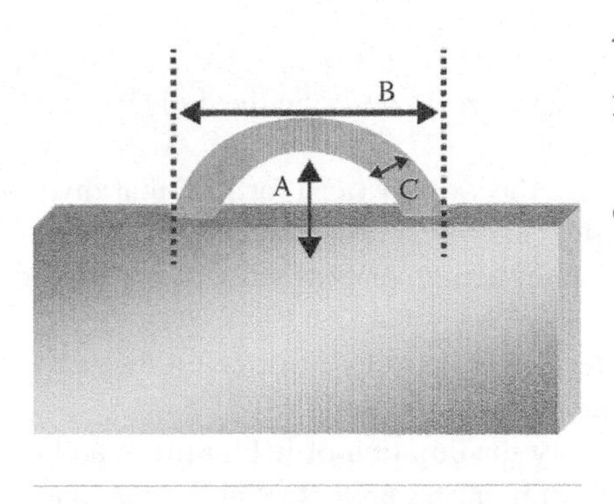

A Internal width = 6.4 cm

B External width: for one hand = 12 cm
for 2 hands = 24 cm

C Diameter: as a function of the weight of the load:

C(cm)	Weight (kg)
0.6	under 9
1.3	7–9
2	over 9

FIGURE 4.10 Description of a good grasp.

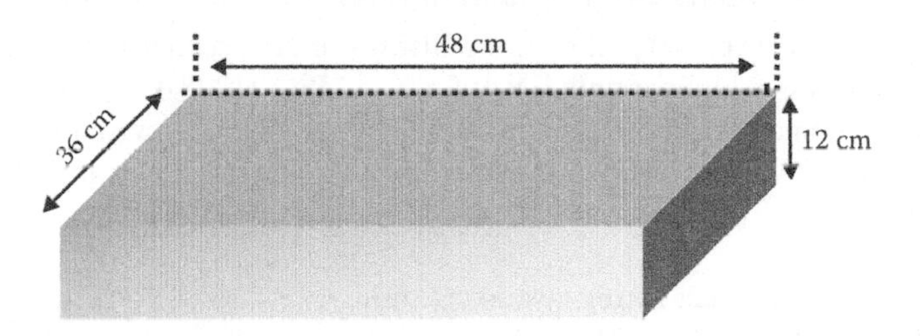

FIGURE 4.11 Maximum measurements recommended for an object to be manually transported.

Figure 4.10 schematically summarises an optimal coupling: It is worth noting that the ideal gripping spaces are geometrically determined also by the weight of the object to be lifted. For instance, an optimal handle design has (Figure 4.10):

- A diameter of at least 2 cm for lifting loads weighing more than 9 kg (C)
- 12 cm length (B)
- 6 cm clearance (A)
- Cylindrical or elliptical shape
- A smooth, nonslip surface

Based on the aforesaid definitions, a coupling is optimal ($CM = 1$) when all the criteria described are met. Depending on the type of coupling and the shape of the object, CM may decrease to poor ($CM = 0.9$).

Considering the difficulty in distinguishing a fair coupling from a poor one, it is advisable to use only good coupling multipliers ($CM = 1$) with the grip described in Figure 4.10, and to class all other couplings as poor ($CM = 0.9$).

Figure 4.11 describes the maximum measurements recommended by the literature for an object to be carried manually without hampering walking or vision. These measurements are indicative and are not based on specific regulations.

4.4.6 Frequency Multiplier (FM)

To define the frequency multiplier it is necessary first to describe the daily work pattern (Figure 4.1).

This first basic analysis mainly focuses on the work pattern for jobs that involve both the manual handling of loads, including carrying, pulling, or pushing (*manual handling times*), and light work without manual lifting, or consisting of pauses, i.e., *recovery periods.*

The description of the daily work pattern also involves identifying the way in which loads are handled and the relative quantities lifted in a shift.

This preliminary analysis serves to identify the duration of lifting times and the frequency with which loads are handled: Both numbers are essential for calculating the FM.

4.4.6.1 Calculating Lifting Duration According to the Revised NIOSH Lifting Equation (RNLE)

Table 4.3 indicates the criteria for defining lifting duration scenarios. The criteria set out in this table are compliant with EN 1005-2 and ISO 11228-1.

Recently, Thomas Waters, the main author of the RNLE, revised the definition of short duration [Waters, 2006] according to the following: Short duration is considered when:

1. The duration of each lifting task is ≤60 min.
2. Lifting work is followed by other light work activities with no lifting, or pauses in duration of ≥100%.

This new definition of short duration is obtained when the recovery period is at least as long as the previous lifting activities (no longer than 1 h continuously).

This update is of considerable practical relevance, especially in light of the possibility of reducing risk by rotating tasks. For example, a short duration can be obtained by simply alternating manual lifting tasks with jobs that do not

TABLE 4.3
Definition of Three Levels of Duration

A. Short duration, if conditions A1 and A2 = TRUE, where:
 A1. Each lifting task ≤ 60 min
 A2. Each block (break or light work) following lifting task ≥ 100% (time) of lifting task
B. Medium duration, if conditions B1, B2, and B3 = true, where:
 B1. Not short duration
 B2. Each lifting task ≤ 120 min
 B3. Each block (break or light work) following lifting task ≥ 30% (time) of lifting task
C. Long duration, if conditions C1 and C2 = true, where:
 C1. Not short duration
 C2. Not medium duration

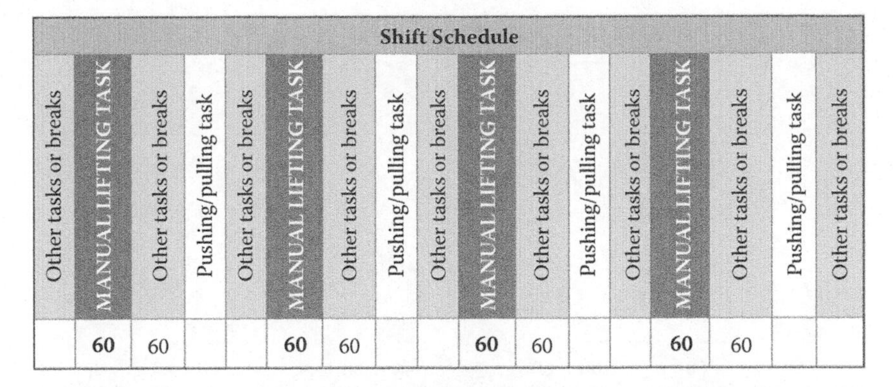

(a) Short duration

Shift Schedule																
Other tasks or breaks	MANUAL LIFTING TASK	Other tasks or breaks	Pushing/pulling task	Other tasks or breaks	MANUAL LIFTING TASK	Other tasks or breaks	Pushing/pulling task	Other tasks or breaks	MANUAL LIFTING TASK	Other tasks or breaks	Pushing/pulling task	Other tasks or breaks	MANUAL LIFTING TASK	Other tasks or breaks	Pushing/pulling task	Other tasks or breaks
	120	60			**60**	60			**40**	60			**20**	60		

(b) Medium duration

Shift Schedule																
Other tasks or breaks	MANUAL LIFTING TASK	Other tasks or breaks	Pushing/pulling task	Other tasks or breaks	MANUAL LIFTING TASK	Other tasks or breaks	Pushing/pulling task	Other tasks or breaks	MANUAL LIFTING TASK	Other tasks or breaks	Pushing/pulling task	Other tasks or breaks	MANUAL LIFTING TASK	Other tasks or breaks	Pushing/pulling task	Other tasks or breaks
	120	30			**60**	60			**30**	90			**30**	60		

(c) Long duration

FIGURE 4.12 Examples of short, medium, and long shift schedules.

involve lifting every hour (visual inspection jobs or recovery times or light work without manual lifting, pulling, or pushing). A few applications are shown in Figure 4.12.

Shift Schedule																
Other tasks or breaks	MANUAL LIFTING TASK	Other tasks or breaks	Pushing/pulling task	Other tasks or breaks	MANUAL LIFTING TASK	Other tasks or breaks	Pushing/pulling task	Other tasks or breaks	MANUAL LIFTING TASK	Other tasks or breaks	Pushing/pulling task	Other tasks or breaks	MANUAL LIFTING TASK	Other tasks or breaks	Pushing/pulling task	Other tasks or breaks
	60		60		**60**	60			**60**	60			**60**	60		

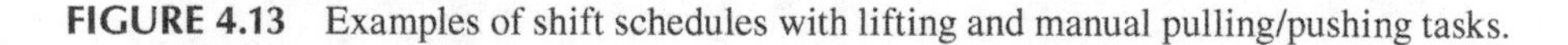

FIGURE 4.13 Examples of shift schedules with lifting and manual pulling/pushing tasks.

Figure 4.12 contains several details that are worth noting:

1. In all three scenarios, the total net duration of manual lifting tasks is 240 min during the shift.
2. In Example B, although the shift includes other short duration manual lifting periods, it is classified as medium duration because it includes one continuous lifting session of 120 min followed by 60 min of no lifting.
3. In Example C, the 30 min of no lifting is presumed not to allow enough recovery time for the 120 min of continuous lifting. This duration is thus classified as long (120 min plus the following 60 min, since the 30 min recovery time between the two lifting sessions is insufficient).

When the shift includes pulling/pushing tasks, these cannot be counted as recovery times: The duration of such tasks must be added to the lifting sessions (Figure 4.13). In the example, the duration is regarded as long because it consists of a first lifting session (60 min), then an intermediate pulling and pushing period (60 min), followed by another lifting session (60 min) totaling 180 min of manual handling with no recovery time.

4.4.6.2 Calculating Frequency

To calculate lifting frequency, the original NIOSH manual [Waters et al., 1994] suggested observing the number of lifts per minute over a period of at least 15 min, counting the number of pieces lifted during the observation period and then calculating the frequency per minute. However, this method could over- or underestimate frequency, depending on the pace of the worker during the observation periods.

Conversely, using the data obtained from the *shift schedule* proposed previously (Figure 4.1) allows the data to be more objective and closer to actual working conditions.

Indeed, by considering and adding the different periods devoted to lifting tasks during the shift, it is possible to obtain the first parameter for calculating frequency, i.e., *the total amount of time engaged in lifting activities (duration) per shift.*

Once the total number of lifts (or total number of objects to be lifted) per shift is known (based on production or commercial data), it is easy to calculate the frequency using the following equation:

Frequency = Number of lifts per shift/Duration of manual lifting per shift (min) (4.6)

4.4.6.3 Calculating Frequency Multiplier

Once the duration and frequency have been determined, it is possible to choose the corresponding frequency multiplier (FM) using Table 4.4.

The frequency multiplier, like the other multipliers, will have a maximum value of $FM = 1$, dropping to $FM = 0$ depending on the individual frequency and work duration values measured.

For frequencies below the minimum frequency shown in Table 4.4 (0.1 lifts/min), the risk evaluation must still be computed based on a frequency multiplier of $FM = 1$.

Table 4.4 shows two groups of multipliers, one for lifts with the vertical height (V) of the hands at the origin at less than 75 cm from the floor, and one at more than 75 cm.

TABLE 4.4
Frequency Multiplier Table (FM)

Frequency (lift/min)	Work Duration			Frequency (lift/min)	Work Duration		
	≤1 h	>1 but ≤2 h	>2 but ≤8 h		≤1 h	>1 but ≤2 h	>2 but ≤8 h
<0.1	1	1	1	≤0.1	1	1	1
0.1–0.2	1	0.95	0.85	0.1–0.2	1	0.95	0.85
0.5	0.97	0.92	0.81	0.5	0.97	0.92	0.81
1	0.94	0.88	0.75	1	0.94	0.88	0.75
2	0.91	0.84	0.65	2	0.91	0.84	0.65
3	0.88	0.79	0.55	3	0.88	0.79	0.55
4	0.84	0.72	0.45	4	0.84	0.72	0.45
5	0.80	0.60	0.35	5	0.80	0.60	0.35
6	0.75	0.50	0.27	6	0.75	0.50	0.27
7	0.70	0.42	0.22	7	0.70	0.42	0.22
8	0.60	0.35	0.18	8	0.60	0.35	0.18
9	0.52	0.30	0.00	9	0.52	0.30	0.15
10	0.45	0.26	0.00	10	0.45	0.26	0.13
11	0.41	0.00	0.00	11	0.41	0.23	0.00
12	0.37	0.00	0.00	12	0.37	0.21	0.00
13	0.00	0.00	0.00	13	0.34	0.00	0.00
14	0.00	0.00	0.00	14	0.31	0.00	0.00
15	0.00	0.00	0.00	15	0.28	0.00	0.00
>15	0.00	0.00	0.00	>15	0.00	0.00	0.00
$V < 75$ cm				$V \geq 75$ cm			

The difference between the values shown in the two tables is minimal and restricted to very high frequencies: To simplify the assessment, it is advisable to use the most protective value, i.e., the value that gives the lowest multiplier (corresponding to height of hands less than 75 cm from the floor).

The software programs illustrated later also use only this approach for estimating the FM.

4.4.6.4 Examples of How to Determine the FM Value

Below are examples of how to determine the FM in particular situations in which, with a constant frequency and total lifting duration per shift, the FM depends on the distribution of the work time with manual handling in the shift.

Analysis of a Task with a Total Lifting Duration of 240 min and a Total of 600 Lifts
The frequency is 2.50 lifts/min. Depending on how the lifting tasks are distributed it is possible to depict three different scenarios determining three different FMs (Figures 4.14 to 4.16):

1. Continuous duration of lifting tasks: Long duration. In this example, crossing duration (long) with frequency produces an FM of 0.61 (Figure 4.14).
2. Noncontinuous lifting duration: 120 min of manual lifting interrupted by a sufficiently long recovery period or other light work activity (with no load lifting). In this example, crossing duration (medium) with frequency produces an FM of 0.82 (Figure 4.15).
3. Noncontinuous lifting duration: 60 min of manual lifting interrupted by 60 min of recovery time or other light work activity (with no load lifting). In this example, crossing duration (short) with frequency produces an FM of 0.90 (Figure 4.16).

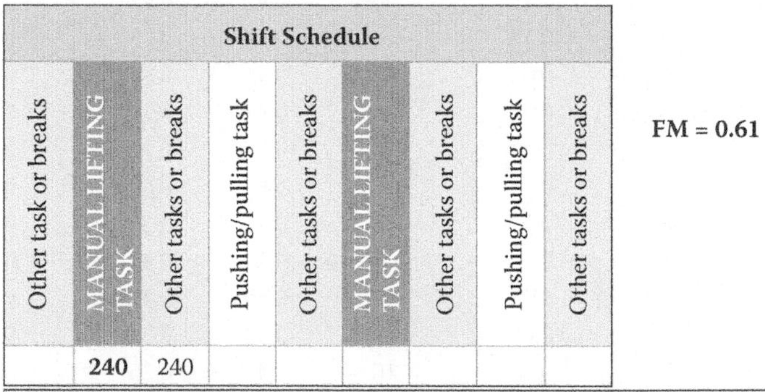

Shift Schedule								
Other task or breaks	MANUAL LIFTING TASK	Other tasks or breaks	Pushing/pulling task	Other tasks or breaks	MANUAL LIFTING TASK	Other tasks or breaks	Pushing/pulling task	Other tasks or breaks
	240	240						

FM = 0.61

Long Duration	
Shift Duration (min.):	480
MMH Duration (min.):	240
Lifting Frequency (actions per min.):	2.5

FIGURE 4.14 Example of determination of FM in a long duration task involving lifting 600 objects (total duration of lifting in the shift: 240 min, frequency: 2.5 lifts/min).

Shift Schedule								
Other tasks or breaks	MANUAL LIFTING TASK	Other tasks or breaks	Pushing/pulling task	Other tasks or breaks	MANUAL LIFTING TASK	Other tasks or breaks	Pushing/pulling task	Other tasks or breaks
	120	120			120	120		

FM = 0.82

Medium Duration	
Shift Duration (min.):	480
MMH Duration (min.):	240
Lifting Frequency (actions per min.):	2.5

FIGURE 4.15 Example of determination of FM in a medium duration task involving lifting 600 objects (total duration of lifting in the shift: 240 min, frequency: 2.5 lifts/min).

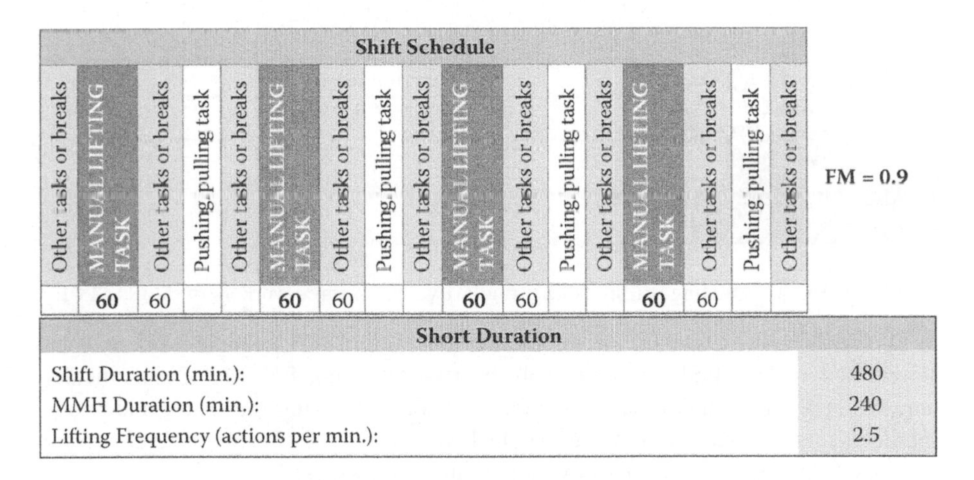

FIGURE 4.16 Example of determination of FM in a short duration task involving lifting 600 objects (total duration of lifting in the shift: 240 min, frequency: 2.5 lifts/min).

4.4.7 Person Multiplier (PM): Lifting by Two or More Workers According to EN 1005-2 and ISO 11228-1

While the original NIOSH formula does not include additional multipliers for lifting tasks performed by two or more workers (team lifting), both EN 1005-2 and ISO 11228-1 envisage adjustments to lifting indexes when the lifting task is performed by two or more workers simultaneously.

Although the two standards adopt different mathematical approaches, they are comparable, and both stress the need to introduce factors reducing the recommended mass (or RWL), when the lifting task takes place with these procedures.

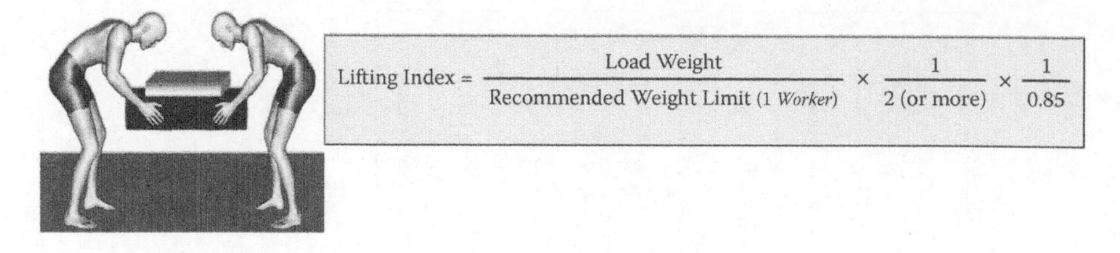

FIGURE 4.17 Calculating lifting index using PM value for two workers in EN 1005-2.

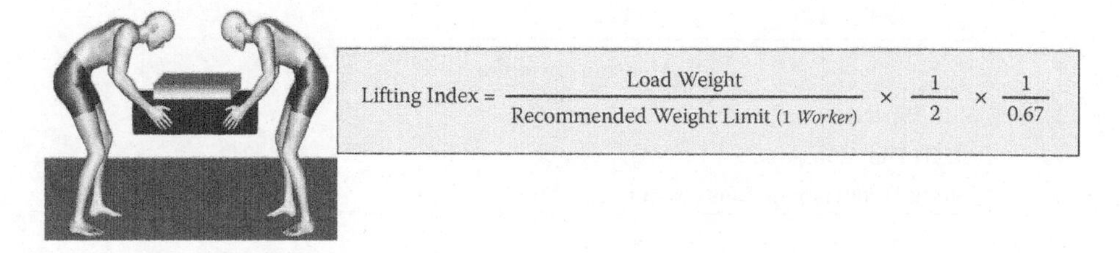

FIGURE 4.18 Calculating lifting index using PM value for two workers in ISO 11228-1.

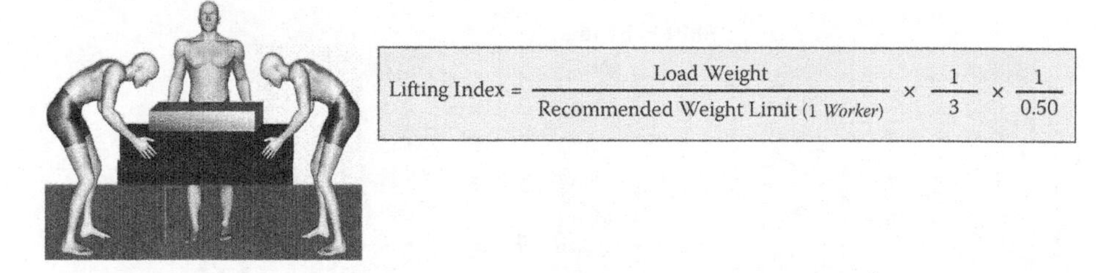

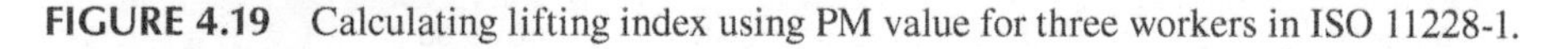

FIGURE 4.19 Calculating lifting index using PM value for three workers in ISO 11228-1.

Specifically, to determine the recommended weight (for a single worker when a weight is to be lifted simultaneously by two workers), EN 1005-2 specifically requires applying a multiplier ($PM = 0.85$); when computing the lifting index one should take consequently half the weight lifted (Figure 4.17). No differences are envisaged when the load is lifted by more than two workers.

ISO 11228-1 only describes the case where there are two or three workers simultaneously lifting a load:

1. For two workers the standard states that "the lifting capability of 2 people is around two thirds of the sum of their individual capabilities." Translated into formulas similar to the EN approach, this means halving the total mass lifted and applying a multiplier ($PM = 0.67$) when computing the individual (for one member of the team) recommended weight (Figure 4.18).
2. For three workers the standard states that "the lifting capability of 3 people is around half of the sum of their individual capabilities." Translated into formulas, this means dividing by 3 the total mass lifted and applying a multiplier ($PM = 0.50$) when computing the individual (for one member of the team) recommended weight (Figure 4.19).

In short, when the lifting task is performed by two or more workers, one may use the different approaches derived from the two regulations, considering that they lead to some differences (from concrete applications about a 10% difference between the two approaches resulted) in the final lifting index.

EN 1005-2, which is compulsory for machinery design in Europe, determines risk using a slightly less conservative approach to lifting actions performed by two or more workers but suggests directly a simpler, more practical approach rather than the ISO. The software presented in this book uses the CEN EN 1005-2 solution. However, it should be noted in general that in order for the formulas (or the suggested multipliers) to apply, each worker in the team (lifter) must have sufficient coupling, an unrestricted posture, and the lifters must not interfere with each other's movements during the course of the lift.

4.4.8 One-Hand Multiplier (OM): One-Hand Lifting According to EN 1005-2 and ISO 11228-1

The two standards treat one-hand lifting differently. Indeed while EN 1005-2 proposes a specific multiplier for computing RWL, ISO 11228-1 (and the original RNLE) simply states that the formula for the final recommended weight (and lifting index) only computes lifting conditions for workers using both upper arms.

Hence, to estimate risk in a situation where the object is lifted with just one hand, the evaluation method described in EN 1005-2 could be used.

This regulation applies an additional multiplier factor ($OM = 0.6$) to the formula for determining the recommended weight limit (Figure 4.20).

If the worker lifts one object with each hand (i.e., one, named load 1, with the left hand and one, named load 2, with the right), the following procedure should be used to estimate the final lifting index:

- Check that the sum total of the two weights is not higher than the reference mass or load constant (eventually taking gender and age into account).
- Check if the difference between the weight of loads 1 and 2 is greater than 20%. In such case:
  - Compute the lifting index for both limbs relative to the weight lifted with each hand, applying the OM (0.6) for computing the recommended weight.
  - Choose the worst value (the higher LI) as the one representative of risk.
- Otherwise, compute the lifting index taking load weight equal to the sum total of the two weights, as lifting with two hands.

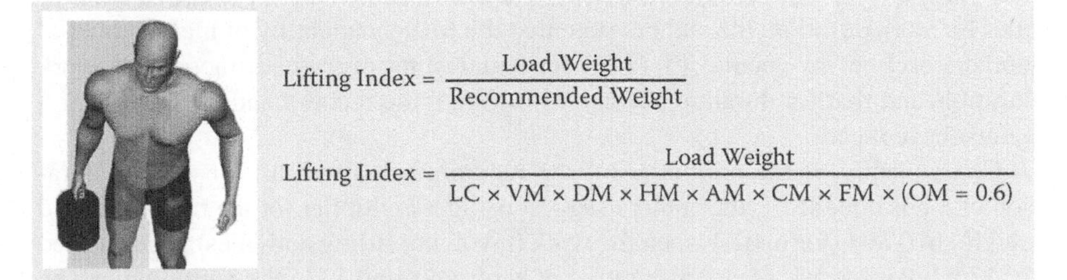

FIGURE 4.20 Calculating lifting index using OM value in EN 1005-2.

4.4.9 Other Conditions and (Suggested) Multipliers

4.4.9.1 Work Duration

Both the original RNLE and the international standards (CEN and ISO) that used it as a reference method consider only a work duration with manual materials handling (MMH) of up to 8 h daily.

In defining frequency multipliers (FMs) as a function of duration, the RNLE was based to some extent on psychophysical criteria, but more significantly on physiologic criteria (i.e., energy expenditure). In this respect, taking 9.5 kcal/min as the maximum reference aerobic capacity (baseline value), the RNLE stated that:

- Repetitive lifting tasks lasting 1 h or less should not require workers to exceed 50% of the baseline value.
- Repetitive lifting tasks lasting 1–2 h should not require workers to exceed 40% of the baseline value.
- Repetitive lifting tasks lasting 2 to 8 h or less should not require workers to exceed 33% of the baseline value.
- No limits were provided for lifting tasks lasting more than 8 h.

Based on these criteria and after numerous adjustments, frequency multipliers (FMs) were eventually calculated as a function of the daily work (and frequency) scenarios for manual lifting activities (see Section 4.4.6) that influence the RWL calculation. In establishing FMs for long duration reference was made to a standard workday of a 480 min shift length (420 min of MMH work and 60 min of breaks or other). However, work, including MMH, may in certain cases last longer than 8 straight h. This commonly happens in the building construction industry, or in supermarkets.

On the other hand, Chapter 1 states that in general, a workday of over 8 h (or over 45 h per week) tends to be associated with a greater occurrence of lower back pain in workers exposed to this condition. For such situations, it might be useful to have criteria (or better still, a multiplier) able to establish an RWL also as a function of a workday lasting even 10–12 h.

In the literature there are some general remarks, but very few specific recommendations. The latter include the suggestions contained in [Mital et al., 1993], which largely concerns energy expenditure limits based on the scientific literature, and ultimately establishes that the energy expenditure limits (for males) for a workday of 10 h should be around 26% of the maximum capacity (3.6 kcal/min), and for 12 h, around 23% of the maximum capacity (3.2 kcal/min). Going on to define a multiplier for work duration, the authors state that the lifting capability of male industrial workers declines by about 3.4% (2.0% for females) for every hour increase in work duration, and that for duration beyond 8 working h the recommended weights for 8 h must be reduced.

Consequently, if the multiplier (of the recommended weight) for a work duration of 8 h is equal to 1, the authors suggest using a multiplier (of the recommended weight) of 0.864 (for males) when the work (involving lifting activities) is performed for 12 h; for intermediate work duration of between 8 and 12 h, the multiplier can be calculated accordingly at values of between 1 and 0.864.

TABLE 4.5
Suggested Extended Time Multiplier (ETM)

Hours (with MMH) in the Shift	≤8	8–9	9–10	10–11	11–12
ETM (Extended Time Multiplier)	1	0.97	0.93	0.89	0.85

Note: Additionally apply to RWL (computed by RNLE) when frequency is ≥0.2 lifts/min and duration scenario is long.

Although this somewhat oversimplified approach assumes a linear trend for a phenomenon (such as muscle fatigue) that is far more complex over time, the authors' suggestion could be adopted, and adapted, to the traditional RNLE to assess the recommended weight limit in cases where lifting activities are regularly performed for prolonged periods of over 8 h of net duration of MMH.

In practice, if the duration scenario has been established (according to the usual criteria) as long and the lifting frequency is ≥0.2 lifts/min, and if the lifting activities are prolonged for over 8 h, the traditional RNLE calculation of the RWL could incorporate an additional extended time multiplier (ETM) as shown in Table 4.5.

Another consideration of the authors, to be used as hypotheses for future studies, is to better modulate the risk scores in the MMH tasks when long duration is present. In fact, currently the mathematical model to calculate the LI gives the same multiplier for a little over 2 h to 8 h. Our proposal would be to introduce, only in the presence of long duration, different multipliers that can modulate the risk, varying the net exposure to MMH hour by hour. For example, for 7–8 h of net MMH, the multiplier is 1, for 6 h it could be 1.02; for 5 h, 1.04; for 4 h, 1.06; etc.

4.4.9.2 Heat Stress Multiplier

Both the original RNLE and the international standards (CEN and ISO) that used it as a reference method consider only the work with MMH in an environment with moderate temperatures.

The authors of the RNLE state that the equation is not applicable to environments where temperatures are unfavourable (i.e., where the air temperature is significantly higher than the range of 19–26°C and the relative humidity is significantly different from the range of 35–50%).

ISO recommends applying ISO 7730 for thermal comfort requirements and states that "extra care should be taken if work has to be done at extremes of temperature; for example, high temperatures or humidity can cause rapid fatigue." However, in certain cases, lifting is performed in hot or humid climatic conditions: In such cases, it is advisable to consider reducing workers' capacities and ultimately lowering the RWLs computed using the traditional RNLE.

In this respect, another useful tip comes from [Mital et al., 1993]. The authors propose using a heat stress multiplier (HSM), based on the wet bulb globe temperature (WBGT) thermal stress index, applied to the recommended weight calculation. The HSM is 1 for WBGT values up to 27°C and 0.88 for WBGT values equal to 32°C. For intermediate WBGT values, the recommended HSM values can be interpolated in parallel.

There are no suggestions for MMH activities in low-temperature conditions.

4.4.9.3 Other Special Conditions

Other special conditions (e.g., working in narrow spaces, bad coupling between the feet and the floor, asymmetry or instability of the load centre of gravity, limited clearance when placing loads) are not specifically mentioned in the original RNLE or in the ISO and CEN standards.

However, several suggestions and useful multipliers can be found in the literature that take unusual conditions into account, and may supplement the RNLE in estimating RWL. For instance, [Mital et al., 1993] provides criteria and specific multipliers for limited headroom, load asymmetry (frontal plane), and load placement clearance.

4.5 LIFTING INDEX AND CONSEQUENT ACTIONS

A synthetic exposure indicator (lifting index) can now be derived from the ratio of the actual weight lifted to the recommended weight for that task in the specific work context. Based on the outcome and for an indicative interpretation of the resulting lifting index (also for better addressing the intervention priorities) one can refer to Table 4.6.

TABLE 4.6
Significance of Lifting Index and Consequent Actions

Lifting Index (LI)	Exposure Level	Interpretation	Consequences
$LI \leq 0.85$ (green)	Acceptable	Exposure acceptable for the majority of the working population (taking gender and age into account).	No consequences.
$0.85 < LI < 1.0$ (yellow)	Borderline	Exposure acceptable for the majority of the working population. However, a sizable percentage of workers could be exposed to very low-level risks.	If possible, improve structural multipliers or adopt different organisational measures. Training workers could be beneficial.
$1.0 < LI < 2.0$ (light red)	Present; low level	A sizable percentage of workers could be exposed to low-level risks.	Redesign tasks and workplaces according to priorities. Train workers and activate health surveillance.
$2.0 < LI < 3.0$ (medium red)	Present; moderate level	An increased percentage of the working population could be exposed to a significant level of risk.	Redesign tasks and workplaces as soon as possible. Train workers and activate health surveillance.
$LI \geq 3.0$ (dark red or purple)	Present; high level	Absolutely unacceptable for the majority of the working population.	Redesign tasks and workplaces *immediately*. Train workers and activate health surveillance.

Table 4.6 was outlined with reference to international standards (i.e., EN 1005-2) and relevant scientific literature on the matter. Actions are required even with a lifting index of 1 to 3. Actions should be undertaken on the basis of risk priorities. Lifting indexes should be reviewed after each action is undertaken. The health of exposed workers should be periodically monitored depending on risk levels (see Chapter 12) and local legislation.

5 Computing the Lifting Index (LI) for Evaluating Single Tasks: Procedures and Examples

5.1 FIRST-LEVEL SCREENING

An initial quick evaluation can be performed without employing computations and formulas, by simply examining whether certain basic assumptions and requirements are met in two different areas:

1. Critical conditions:
   - The existence of critical conditions relating to the work environment or load characteristics (Figure 5.1)
   - The existence of critical conditions relating to the layout of the workstation and the existence of loads heavier than recommended weight limits: critical code (Figure 5.2)
2. Acceptable conditions:
   - Compliance with recommended weight limits in relation to lifting frequency and lifting postures: green code (Figure 5.3)

Having applied the procedure, if all the acceptability criteria are met and there are no critical codes, the condition is defined as *acceptable* and no further evaluations are required. Otherwise, a more detailed evaluation is required using the equations provided.

5.2 EVALUATION OF THE LIFTING INDEX (LI)

After performing the quick evaluation described above, if the task is found to require a full evaluation, this should be performed as described in Figure 5.4, beginning with collecting the data required to assess the RWL and the lifting index.

Here is an example of the calculation.

Figure 5.5 sets out the data concerning the organisation and layout of the task required to calculate the risk associated with a given lifting task (Exercise 1).

For a faster evaluation, the procedure shown in Figure 5.6 can be used (with the sample calculation relative to Exercise 1 to easily compute the risk index associated with single lifting tasks). In the exercise an adult male lifter (load constant = 25 kg, or if one prefers, 23 kg) was considered.

Critical Conditions

Is the working environment unfavourable for manual lifting and carrying?

Presence of extreme (low or high) temperature	No	Yes
Presence of slippery, uneven, unstable floor	No	Yes
Presence and use of stairs	No	Yes
Presence of insufficient space for lifting and carrying	No	Yes

Are there unfavourable object characteristics for manual lifting and carrying?

The size of object reduces the operator's view and hinders movement	No	Yes
The load centre of gravity is not stable (e.g., liquids, items moving around inside of object)	No	Yes
The object shape/configuration presents sharp edges, surfaces, or protrusions	No	Yes
The contact surfaces are too cold or too hot	No	Yes

FIGURE 5.1 Critical conditions relating to the work environment or load characteristics.

Quick Assessment: Critical Conditions

If any one of these keys is present, the risk is *very high* and it is necessary to proceed immediately for redesigning tasks.

Vertical location (origin or destination)	Hands at the beginning/end of the manual lifting, higher than 175 cm or lower than 0 cm	Yes	No
Vertical travel distance	The vertical distance between the origin and the destination of the lifted object is more than 175 cm	Yes	No
Horizontal location	The horizontal distance between the lifted object and the body center of gravity (medium point between the ankles) is more than 63 cm	Yes	No
Asymmetry angle	Asymmetry angle (upper body rotation) more than 135°	Yes	No
Frequency of lifts	More than 15 lifts/min in *short duration* (manual handling lasting no more than 60 min consecutively in the shift, followed by at least 60 min of break or light task)	Yes	No
	More than 12 lifts/min in *medium duration* (manual handling lasting no more than 120 min consecutively in the shift, followed by at least 30 min of break or light task)	Yes	No
	More than 8 lift/min in *long duration* (manual handling lasting more than 120 min consecutively in the shift)	Yes	No

More than the maximum recommended weight for any individual is lifted.

Men (18–45 years old)	25 kg	Yes	No
Women (18–45 years old)	20 kg	Yes	No
Men (<18 or >45 years old)	20 kg	Yes	No
Women (<18 or >45 years old)	15 kg	Yes	No

FIGURE 5.2 Quick assessment: Critical conditions relating to the layout of the workstation and the existence of loads exceeding recommended weight limits: critical code.

Lifting: Quick Assessment of Acceptable Conditions

3–5 kg	Asymmetry (e.g., body rotation, trunk twisting) is absent	No	Yes
	Load is maintained close to the body	No	Yes
	Load vertical displacement is between hips and shoulders	No	Yes
	Maximum permissible frequency: less than 5 lifts/min	No	Yes
5.1–10 kg	Asymmetry (e.g., body rotation, trunk twisting) is absent	No	Yes
	Load is maintained close to the body	No	Yes
	Load vertical displacement is between hips and shoulder	No	Yes
	Maximum permissible frequency: less than 1 lift/min	No	Yes
>10 kg	Loads more than 10 kg are not present	No	Yes

If all the listed conditions are yes, the examined task is acceptable (green code) and it is not necessary to continue the risk evaluation.
If one is no, apply a further detailed evaluation.

FIGURE 5.3 Quick assessment of acceptable conditions.

Operative Steps for Manual Lifting Evaluation

A. Identification of kinds of lifting task(s)	Single or composite or variable or sequential task
B. Description of the workers involved in manual lifting task	Number, homogeneous group
C. Organisational analysis: shift schedule and identification of lifting periods	Evaluation of manual lifting duration
D. Identification of the number of objects manually lifted in a shift	Evaluation of manual lifting frequency
E. Analysis of geometries at the origin and destination of the lifted objects	Study of the layout risk factors

FIGURE 5.4 The operative steps for risk evaluation of manual lifting.

The scheme provides the quantitative values (and qualitative values only for evaluating couplings) for each influencing factor, next to the relative multiplier.

By applying the procedure to all the factors (and multipliers) considered, it is possible to determine the recommended weight limit for each job.

The next step is to compute the weight actually lifted (numerator) vs. the recommended weight limit (denominator) to obtain the lifting index.

This simple calculation scheme is also supplied in dedicated software: The lifting index is obtained by entering the relative load constant (25, 23, 20, or 15 kg considering age and gender of the worker performing the task), the various multipliers, and the weight of the object actually lifted. The software ERGOepm_LI_S automatically calculates the lifting index. The dedicated software (and Exercise 1 performed) can be downloaded free from www.epmresearch.org in English and in Italian, and from www.epminternationschool.org in Spanish.

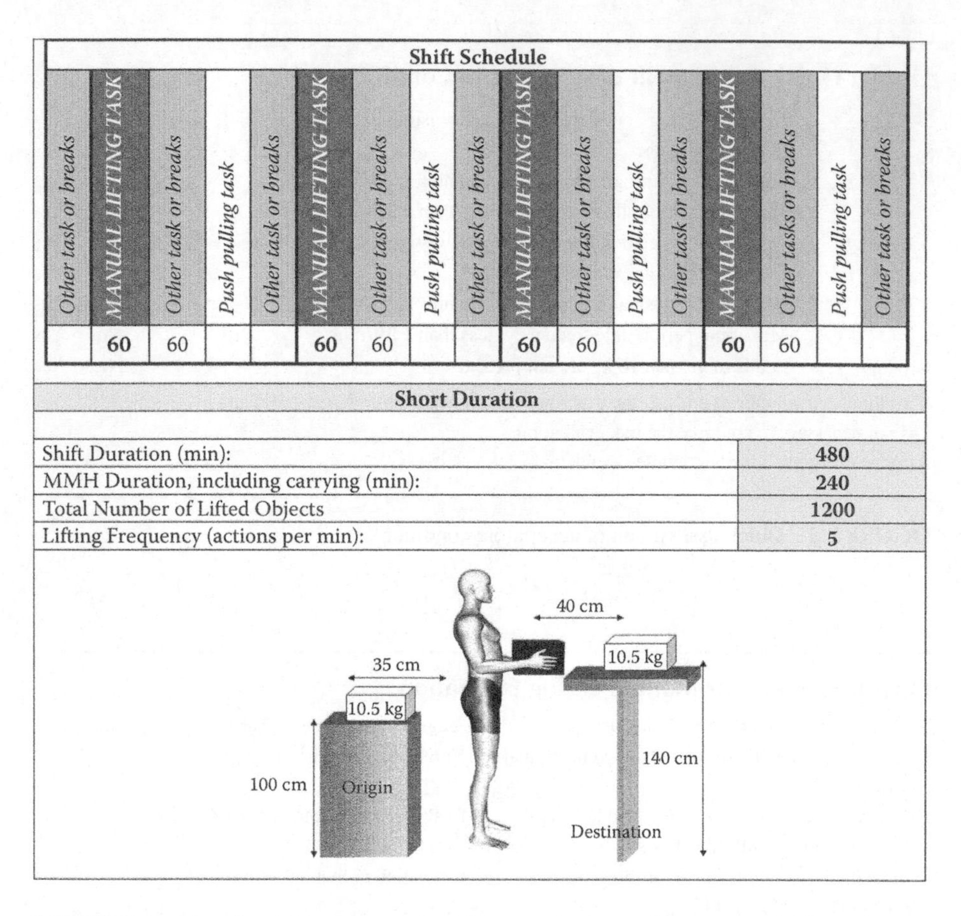

FIGURE 5.5 Exercise 1: Data concerning the organisation and layout required to calculate the LI in a single task.

The example shows two different distances from the body (horizontal locations), one at the origin (35 cm) and the other at the destination (40 cm) of the lift. There are also two different heights of the hands from the floor (vertical locations), one at the origin (100 cm) and the other at the destination (140 cm) of the lift. One might reasonably ask: Which of the two distances from the body and which of the two heights of the hands from the floor should be used?

The NIOSH method proposes calculating:

- A lifting index at the origin
- A lifting index at the destination

The risk is represented by the worst of the two.

Figure 5.6 describes the analysis that led to the worst result (in this case, the destination of the lift).

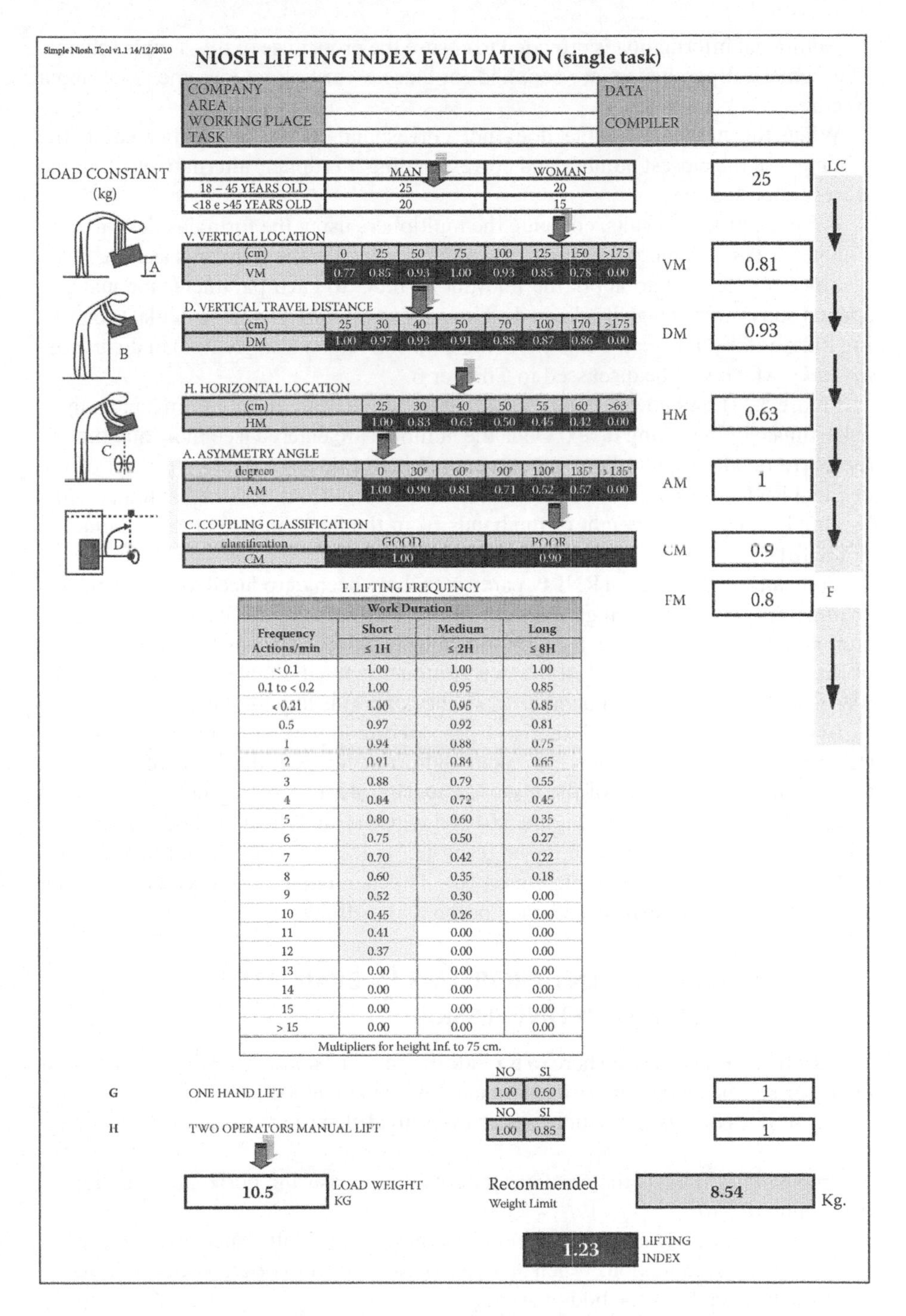

Simple Niosh Tool v1.1 14/12/2010

NIOSH LIFTING INDEX EVALUATION (single task)

COMPANY		DATA	
AREA			
WORKING PLACE		COMPILER	
TASK			

LOAD CONSTANT (kg)

	MAN	WOMAN
18 – 45 YEARS OLD	25	20
<18 e >45 YEARS OLD	20	15

LC: 25

V. VERTICAL LOCATION

(cm)	0	25	50	75	100	125	150	>175
VM	0.77	0.85	0.93	1.00	0.93	0.85	0.78	0.00

VM: 0.81

D. VERTICAL TRAVEL DISTANCE

(cm)	25	30	40	50	70	100	170	>175
DM	1.00	0.97	0.93	0.91	0.88	0.87	0.86	0.00

DM: 0.93

H. HORIZONTAL LOCATION

(cm)	25	30	40	50	55	60	>63
HM	1.00	0.83	0.63	0.50	0.45	0.42	0.00

HM: 0.63

A. ASYMMETRY ANGLE

degrees	0	30°	60°	90°	120°	135°	>135°
AM	1.00	0.90	0.81	0.71	0.52	0.57	0.00

AM: 1

C. COUPLING CLASSIFICATION

classification	GOOD	POOR
CM	1.00	0.90

CM: 0.9

F. LIFTING FREQUENCY

Frequency Actions/min	Work Duration Short ≤ 1H	Medium ≤ 2H	Long ≤ 8H
< 0.1	1.00	1.00	1.00
0.1 to < 0.2	1.00	0.95	0.85
< 0.21	1.00	0.95	0.85
0.5	0.97	0.92	0.81
1	0.94	0.88	0.75
2	0.91	0.84	0.65
3	0.88	0.79	0.55
4	0.84	0.72	0.45
5	0.80	0.60	0.35
6	0.75	0.50	0.27
7	0.70	0.42	0.22
8	0.60	0.35	0.18
9	0.52	0.30	0.00
10	0.45	0.26	0.00
11	0.41	0.00	0.00
12	0.37	0.00	0.00
13	0.00	0.00	0.00
14	0.00	0.00	0.00
15	0.00	0.00	0.00
> 15	0.00	0.00	0.00

Multipliers for height Inf. to 75 cm.

FM: 0.8

		NO	SI	
G	ONE HAND LIFT	1.00	0.60	1
H	TWO OPERATORS MANUAL LIFT	1.00	0.85	1

LOAD WEIGHT KG: 10.5

Recommended Weight Limit: 8.54 Kg.

LIFTING INDEX: 1.23

FIGURE 5.6 Exercise 1. Simple scheme for lifting index evaluation in single lifting tasks (the worst result: the destination).

Additional information is provided to ensure the proper use of this simple scheme. The numerical parameters are supplied for classes varying by increments of about 10 cm.

When the numerical value does not correspond to the one indicated in the scheme, use the closest number and corresponding multiplier; alternatively, find the closest interpolation.

For more precise results, compute the multipliers using the formulas described in Chapter 4. This is a more analytical evaluation, defined in the standards as *third level*.

Since it is tedious to apply the relevant formula to each parameter manually, a special software package has been developed that can analytically calculate the lifting indexes for both *single tasks* (with the data entered on a single line) and *composite tasks*, which will be discussed in Chapter 6.

Figure 5.7 shows how to enter the data into the software for obtaining an analytical estimate of the lifting index. Once the numbers are entered (see the white boxes), the corresponding multipliers appear automatically through the application of the original RNLE formulas. Even the vertical travel distance is obtained automatically, by simply entering the height of the hands from the floor (in cm) at the destination of the lift.

Compared to the original RNLE, parameters have been introduced for lifting by two or more workers, and for single-handed lifting. The software ERGOepm_LI&CLI_C and the performed Exercise 1 can be downloaded free from www.epmresearch.org in English and in Italian, and from www.epminternationalschool.org in Spanish. The files may be updated from time to time, so check periodically for the latest versions.

In the example, the same Exercise 1 proposed in Figure 5.6, both lifting indexes (LI) for origin and destination are calculated: as mentioned, the worst of the two is taken to be representative of the task. Comparing the two lifting indexes obtained using the simplified method (Figure 5.6) and with the analytical method (Figure 5.7) a slight difference can be seen in the result (LI = 1.23 using the simplified method and LI = 1.24 using the analytical method): This negligible difference is due to the different degrees of intrinsic accuracy of the two methods.

5.3 SOLVING FREQUENT PROBLEMS IN EVALUATING MANUAL LOAD LIFTING TASKS

This section has been added here to provide operational solutions to doubts that commonly arise when assessing risks associated with manual load lifting.

The section consists of questions and answers divided into various categories of topics.

Problem 5.1: What is the most accurate way of measuring horizontal location (Figure 5.8)?

Measuring horizontal location is somewhat difficult, especially because the body is in motion. When an object is shifted from origin to destination, its distance from the body varies.

Which horizontal location should be chosen? The answer is the worst one.

How is the horizontal location measured? By keeping the subject immobile in the worst position until the complex measurement is completed.

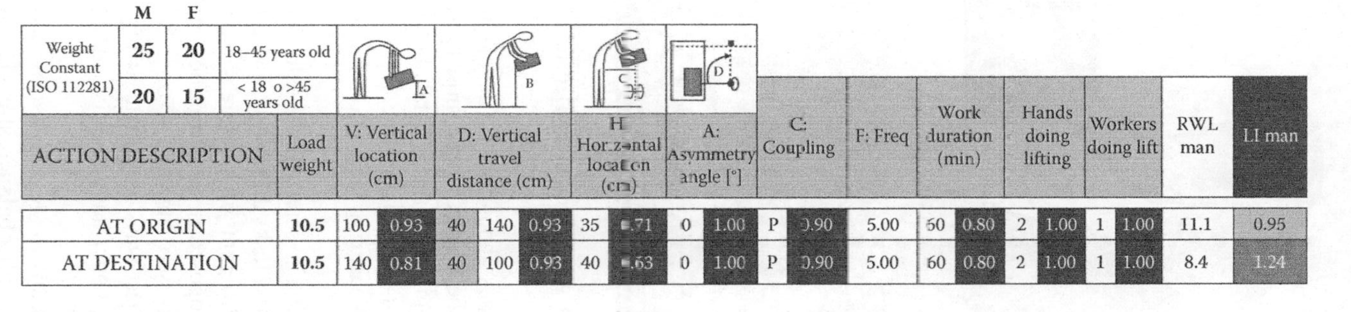

Weight Constant (ISO 112281)	M	F	
	25	20	18–45 years old
	20	15	< 18 o >45 years old

ACTION DESCRIPTION	Load weight	V: Vertical location (cm)		D: Vertical travel distance (cm)			H Horizontal location (cm)		A: Asymmetry angle [°]		C: Coupling		F: Freq	Work duration (min)		Hands doing lifting		Workers doing lift		RWL man	LI man
AT ORIGIN	**10.5**	100	0.93	40	140	0.93	35	0.71	0	1.00	P	0.90	5.00	60	0.80	2	1.00	1	1.00	11.1	0.95
AT DESTINATION	**10.5**	140	0.81	40	100	0.93	40	0.63	0	1.00	P	0.90	5.00	60	0.80	2	1.00	1	1.00	8.4	1.24

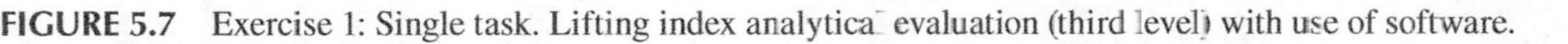

FIGURE 5.7 Exercise 1: Single task. Lifting index analytical evaluation (third level) with use of software.

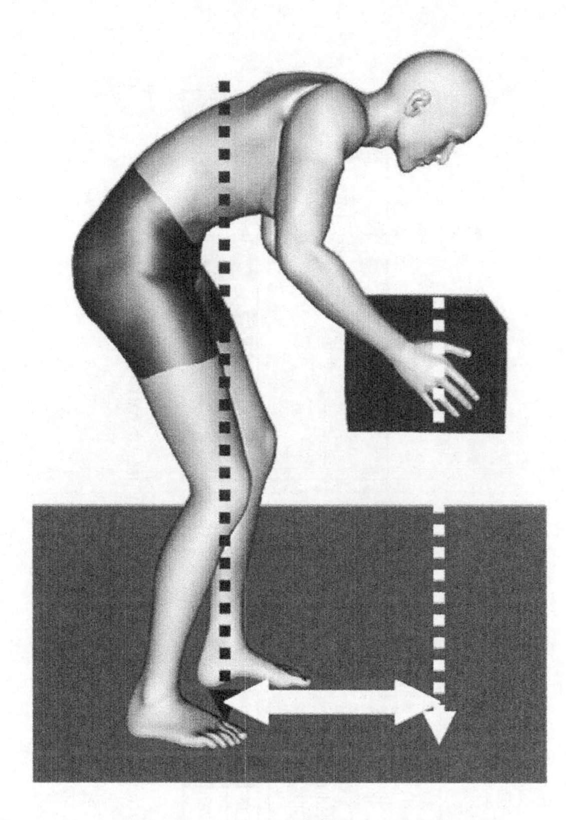

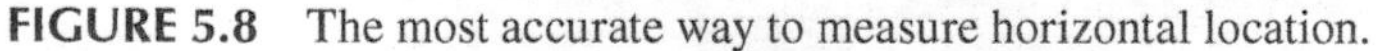

FIGURE 5.8 The most accurate way to measure horizontal location.

First, the worker's back should not be needlessly overloaded: This can be avoided by putting the worker in the most representative position, and removing the object from the worker's hands before measuring. The measurement is performed using a plumb line, a ruler positioned between the worker's ankles, and a measuring tape. Photographic methods of collecting data could be useful.

It is virtually impossible to set a definite value for the horizontal location since workers will perform tasks differently, and even the same worker may perform the same task differently. Therefore it makes more sense to set representative values as proposed in the simplified model for jobs comprised of more than 10 subtasks (see Chapter 7).

Problem 5.2: What is the best way to evaluate lifting when the workers have to lean forward on their upper body to grasp an object that is beyond their reach (Figure 5.9)?

Even if the worker leans the legs, waist, or chest forward to reach the object, without losing his or her balance, stress is still exerted on the intervertebral disks. Therefore the same NIOSH formula is used for the calculation.

If the horizontal location is over 63 cm the condition would be critical.

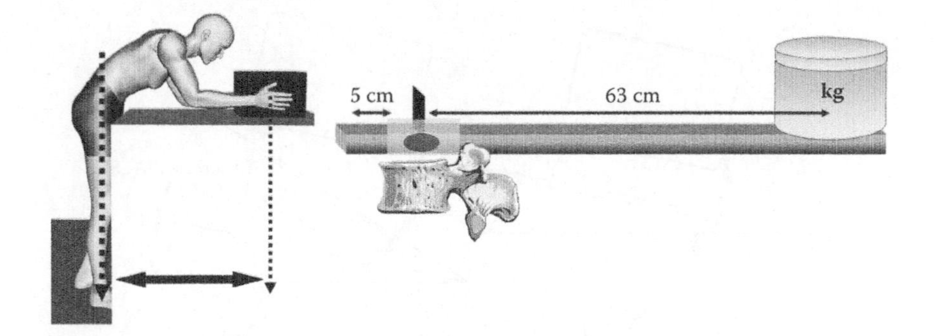

FIGURE 5.9 The best way to evaluate lifting when the workers have to lean forward on their upper body to grasp an object that is beyond their reach.

It is important to note that leaning forward to lift an object that is some distance from the body may generate major stresses on the lumbar spine! Therefore such movements should be avoided.

Problem 5.3: Should a risk assessment be performed for low lifting frequencies (occasional lifting)?

Lifting frequencies of less than 0.1 time/min imply multipliers equal to 1; however, this does not mean that there is no risk, or that the job need not be assessed for risk.

When the objects lifted are very light, task conditions are virtually ideal, and the lifting frequency is low, it is highly unlikely that there is any risk (acceptable condition). However, when the object is significantly heavier, it is important to complete the analysis using the RNLE, even when the frequency multiplier is 1, to assess the impact of multipliers for less than ideal lifting geometries.

Problem 5.4: What is the best way to analyse a task in which the object is first pulled toward the worker and then lifted?

The analysis should be divided into two steps:

Step 1: Use a dynamometer to measure the force required to pull the object, then use the Liberty Mutual tables (or ISO 11228-2 tables) to calculate an exposure index by comparing the actual exerted force (measured) with the corresponding recommended weight (or force).

Step 2: Apply the RNLE to evaluate the lifting task. The horizontal location is measured after the object is pulled closer to the body.

Problem 5.5: Are there differences between men and women when assessing the vertical height of the lift or the vertical dislocation of the load? In other words, does the critical height cutoff of 175 cm also apply to women (Figure 5.10)?

Unfortunately current regulations do not specify different criteria for assessing risk with regard to these two factors.

Stretching up over the head to reach a load determines hyperextension of the lumbar spine, and 175 cm represents the height of the average man. This posture, especially if overloaded with a heavy object, is harmful due to the

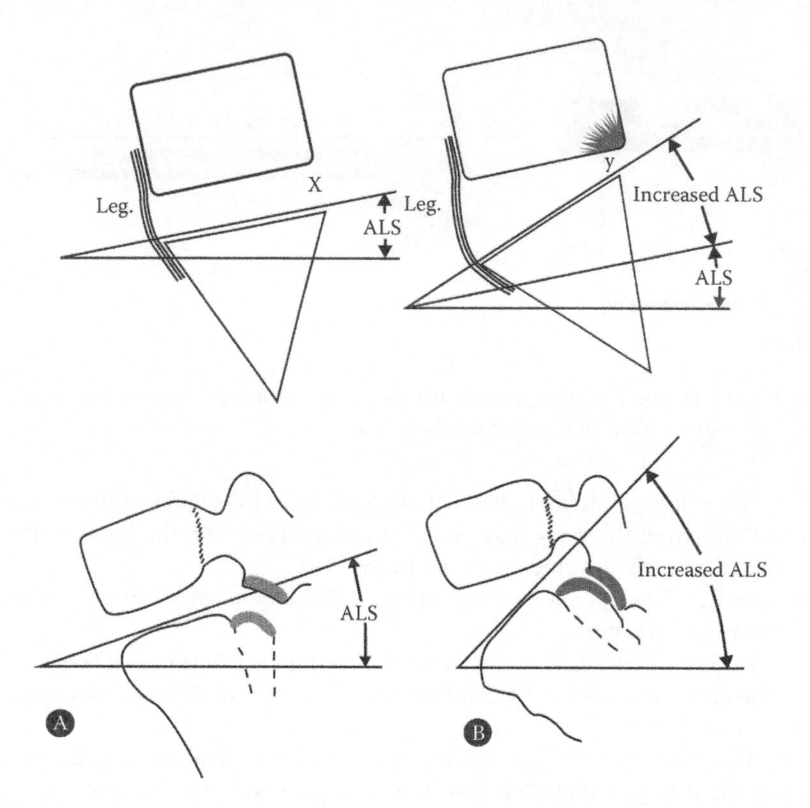

FIGURE 5.10 The biomechanical effects of stretching the lumbosacral spine: widening the anterior vertebral space with excessive stretching of the anterior longitudinal ligament and contraction of the posterior portion, with narrowing of the space between the two joint faces.

biomechanical consequences caused by the increased lumbosacral angle, such as widening the anterior vertebral space with excessive stretching of the anterior longitudinal ligament and contraction of the posterior portion, with narrowing of the joint space.

Even without specific regulations, workers should avoid lifting weights above head level.

Problem 5.6: How should we evaluate the lifting of objects that are thrown to pass them over obstacles or to reach excessive heights before lowering them?

Throwing of objects should be avoided: The RNLE cannot be applied to lifts featuring such operating methods.

Problem 5.7: How should we evaluate the lifting of objects in restricted work spaces that prevent the worker from maintaining a standing position (low overhead clearance)?

ISO 11228-1 asks for "sufficient headroom to avoid stooping postures whilst handling an object." EU legislation on the matter says that "the characteristics of the work place may increase the risk of injury due to biomechanical overload, especially back pain, when the available space, particularly vertical, is insufficient for performing the required task and/or

TABLE 5.1
Indications for Further Reducing the Recommended Weight Limit in Relation to the Degree to Which the Work Space Restricts Standing Upright (Low Overhead Clearance)

Stature	Position 100% Upright	Position 95% Upright	Position 90% Upright	Position 85% Upright	Position 80% Upright
Multiplier	1.00	0.60	0.40	0.38	0.36

the work place or environment does not allow the worker to manually lift loads at a safe height or in a suitable position."

The handling of loads in restricted spaces should therefore be avoided, and any model for evaluating risk applied under such conditions would evidence an extremely high risk. Should it be necessary to make an evaluation, however (for instance, because it was specifically required by the supervisory authorities or for medicolegal purposes), indications can be found in the literature [Mital et al., 1993] for further reducing the recommended weight depending on how restricted the work space is (Table 5.1).

Problem 5.8: Is it possible to evaluate manual lifting of a patient or an animated object using the NIOSH method?

Both ISO 11228, Part 1, and NIOSH clearly state that it is not possible to use the LI approach for these lifting tasks. It should be noted that (within 2012) ISO will produce a specific TR (ISO TR 12296) on the manual lifting of patients, where methods for evaluating (and managing) risks associated with lifting patients will be widely reported: Among these it is worth recalling the MAPO (Movement and Assistance of Hospitalized Patients) method.

However, Thomas Waters [2007] proposed using the LI for partly cooperative patients and under ideal conditions provided the weight lifted should be less than 15.9 kg. For heavier weights, the worker would be at risk and lifting aids should be used. Consequently, if conditions are not ideal (for example, if the patient is noncooperative, the weight of the patient cannot be calculated, the movement is rapid, or the geometries are subject to change), the LI method cannot be used and reference should be made to the ISO TR previously reported.

From a practical standpoint, the LI method could be used for lifting toddlers (up to about 3 years of age), provided they weigh less than 16 kg.

Problem 5.9: How can the horizontal distance multiplier be evaluated when using a long-handled tool or when the object cannot be grasped in the middle? Should the calculation still take the distance between the hand grasp and the body centre (i.e., the midpoint of the line joining the inner ankle bones)?

In some lifting situations, especially involving the use of long-handled tools, the centre of the object does not coincide with the hand grasp: In this case the horizontal distance from the body must be found as the distance from the load centre and the usual midpoint of the line joining the ankle bones.

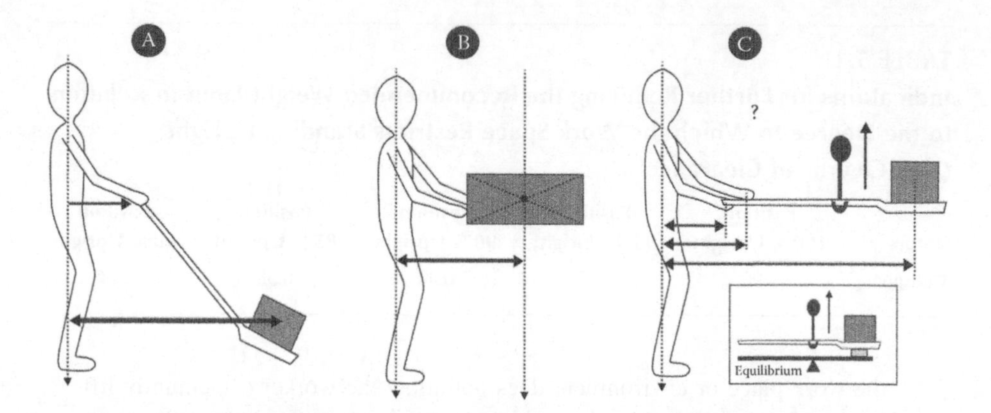

FIGURE 5.11 How to evaluate the horizontal distance multiplier (horizontal distance from the body) when the object cannot be grasped in the middle of the hand grasp.

Consider the following working conditions:

Figure 5.11A: Using a shovel. In this condition it is relatively easy to find the centre of the shovel + load, because it coincides with the centre of the load on the shovel. However, it should be noted that such conditions are clearly excluded from the application of the RNLE and more complex biomechanical models should be used.

Figure 5.11.B: Lifting bulky objects asymmetrically in the figure, the operator cannot grasp the load at its centre. The horizontal distance multiplier is calculated based on the load centre and not the hand grasp.

Figure 5.11C: Lifting a complex tool with two hands placed at different distances. In some conditions it is more difficult to locate the centre of the tool + load. In these cases it is advisable to mechanically locate the central point where the two lever arms (see figure) are balanced (for instance, by simply attaching the tool to hanging scales). The point of equilibrium coincides with the centre, and this can be used to identify the horizontal distance.

Problem 5.10: Is it sufficient to apply the RNLE approach for estimating the risk of biomechanical overload in tasks involving manual load handling?

The answer is definitely no. Often, the manual handling of loads also includes carrying, holding, pulling, pushing, and handling low loads (<3 kg) at high frequencies (overloading the upper limbs), which therefore need to be evaluated using, for instance, the methods and criteria suggested in the relevant technical standards ISO 11228, Part 1 [ISO, 2003], Part 2 [ISO, 2007a], and Part 3 [ISO, 2007b].

Problem 5.11: Why is it that redesigning the heights of shelves on which objects are placed does not always improve lifting indexes?

To explain this, see Table 5.2 highlighting the different relevance of the risk factors determining exposure levels. The first factor is frequency, next comes horizontal location, and so on. The height of the load from the ground has only a medium relevance; therefore redesigning this aspect may fail to produce major reductions in the lifting index.

TABLE 5.2
Different Relevance of Risk Factors Determining Exposure Level

	Multiplier	Degree	Percentage Reduction
Individual characteristics	Female gender	Medium	20%
	Age: <18 or >45 years	Medium	20–25%
Lift type: physical factors	Height of weight from floor	Medium	22%
	Vertical dislocation	Low	14%
	Horizontal distance	High	58%
	Twisting of upper body	High	48%
	Type of grip	Low	10%
	Lifting (with only 1 arm)	High	40%
Organisational factors	Frequency and duration	High	75%
	Lifting (2 operators)	Low	15%

6 Computing the Lifting Index for Evaluating Composite Tasks (CLI): Procedures and Examples

6.1 EVALUATION OF THE COMPOSITE LIFTING INDEX (CLI)

As already described for *single tasks*, if, after performing the quick evaluation described in the preceding chapter, the task requires a full evaluation, then it should be performed as described in Figure 5.4, beginning with collecting the data required to assess the lifting index.

Jobs sometimes call for workers to perform *composite lifting tasks*, i.e., tasks where the weight is the same but it is moved over several geometries (Figure 6.1); in such cases evaluations have to take into account the contribution of various subtasks, in determining risk conditions.

What are subtasks? As shown in Figure 6.2, in Exercise 2, the worker removes objects of the same weight from one of five shelves, and places them on a conveyer belt whose height is constant. All of the five shelves are at the same distance from the body (only shelf A is at a different distance, equal to 45 cm). Therefore, there are five subtasks conveying the concept of the presence of five single tasks. Each single task must first be evaluated separately, and then the composite lifting index (CLI) is estimated; the term *composite* is used because the task is comprised of several single subtasks.

In Exercise 2, the *total frequency* is 12 times/min.

Since there are five subtasks, the intrinsic frequencies with which the objects are lifted in each task are 2.4 times/min (total frequency 12 times/min divided up equally between the five subtasks).

Using a general computation procedure, a lifting index is calculated for each individual subtask:

1. Using the intrinsic frequency (in the example, 2.4 times/min)
2. Using the total frequency (in the example, 12 times/min)

In general, the CLI is determined by the lifting index (LI) for the most overloading subtask, plus an amount determined by the lifting indexes of the other subtasks.

All the numerical layout parameters, described for single tasks, also need to be measured for each subtask (Figure 6.2).

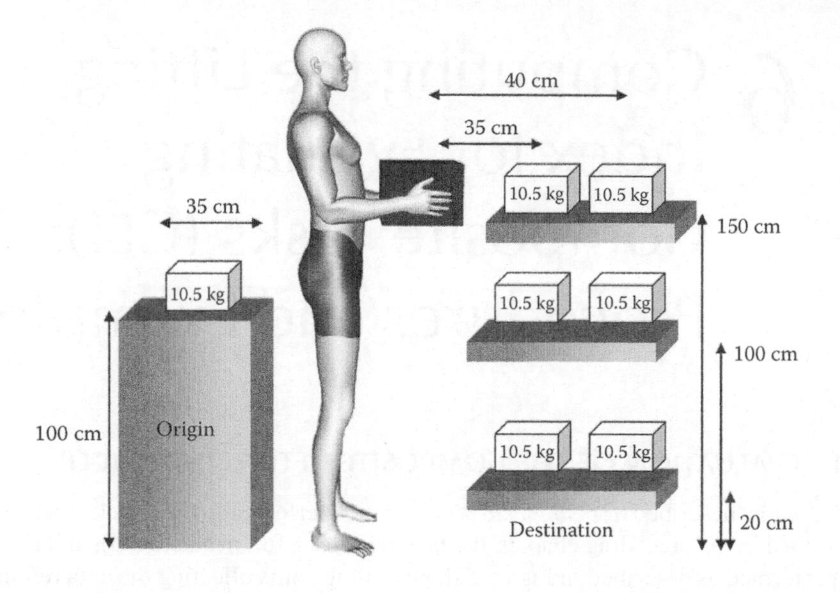

FIGURE 6.1 Example of a composite task: Lifting an object of the same weight over several geometries.

The assumption underlying the calculation of the CLI is supplied by the NIOSH application manual [Waters et al., 1994], which indicates that a composite task may generally be represented by the following formula:

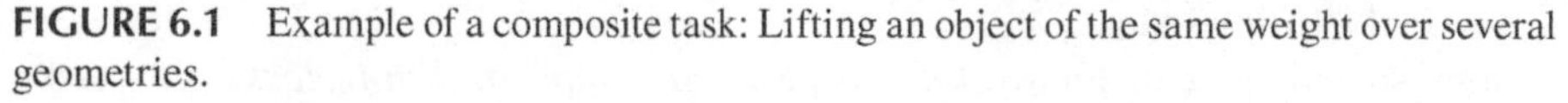

$$CLI = LI1 + \Sigma\Delta LIn \tag{6.1}$$

where $\Sigma\Delta LIn$ is more clearly set out in Figure 6.3.

For a better understanding of the variables that have to be computed, several mathematical steps are also described in Figure 6.3.

A simplified procedure is now provided to facilitate this complex evaluation, which would otherwise unquestionably require the use of software; the procedure supplies an instant snapshot of at least the range that the CLI value will fall in:

1. The lower limit is computed by identifying the single task with the highest LI: In this case, the frequency multiplier will be that of the single subtask.
2. The upper limit is obtained by applying the total frequency multiplier for the whole task to the same most critical subtask.

As mentioned, it is extremely difficult to calculate the CLI formula manually.

Therefore it is advisable to use the same dedicated software model (ERGOepm_LI&CLI_C) recommended for calculating the LI in single tasks, but in analytical mode, available at www.epmresearch.org.

Figure 6.4 provides an example of compiling the software, based on the organisation and layout used in Figure 6.2, Exercise 2. The final results are shown according

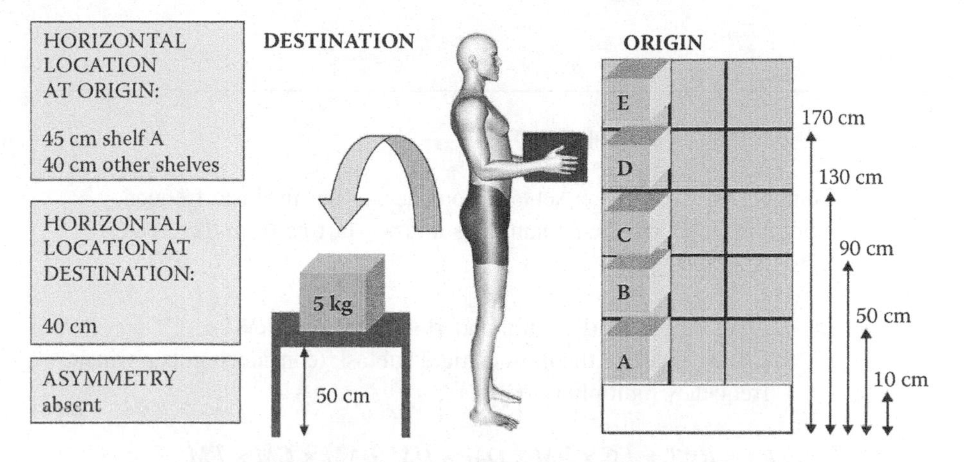

5 HEIGHTS AT ORIGIN AND 1 HEIGHT AT DESTINATION = 5 SUB-TASKS

Subtask	L. Load weight	V. Vertical location	D. Vertical travel distance	H. Horizontal location	A. Asymmetry	Frequency	Duration Min.	C Coupling
A	5	10	40	45	0	2.4	60	Poor
B	5	50	0	40	0	2.4	60	Poor
C	5	90	40	40	0	2.4	60	Poor
D	5	130	80	40	0	2.4	60	Poor
E	5	170	120	40	0	2.4	60	Poor

FIGURE 6.2 Exercise 2: Composite task. Data required to compute CLI.

to reference masses, by gender and age, suggested by ISO 11228-1. Using a reference mass of 23 kg, as in the original RNLE, the final result would be $LI = 1.42$.

Once the numbers are entered for the various subtasks, the corresponding multipliers appear automatically through the application of the original NIOSH formulas. As shown, the software assigns a separate horizontal line to each subtask.

Compared to the original NIOSH formula, parameters have been introduced for lifting by two or more workers, and for single-handed lifting.

6.2 COMPUTING THE NUMBER OF SUBTASKS PRIOR TO EVALUATING THE CLI: EXAMPLES

Once it has been decided that a composite task needs to be assessed, it is crucial, as mentioned earlier, to identify the number of potential subtasks to consider before computing the CLI.

It should be noted that to accurately compute the CLI, there must not be too many subtasks (10 is the maximum). Since the overall frequency has to be divided by the number of subtasks, if there are more than 10 subtasks to consider the final result

1. Lifting index (LI) computation

   Compute an LI for each subtask considering the intrinsic frequency multiplier (FM) and then renumber subtasks starting from the greatest LI to the lowest.

   LI = Load weight (L)/Recommended weight limit (RWL)
   LI_1 = Lifting index of the most critical subtask (considering its intrinsic frequency multiplier—FM_1)

$$RWL = LC \times VM \times DM \times HM \times AM \times CM \times FM$$

   where:
   LC = Load constant
   VM = Vertical location multiplier
   DM = Vertical travel distance multiplier
   HM = Horizontal distance multiplier
   AM = Asymmetry angle multiplier
   CM = Coupling multiplier
   FM = Lifting frequency multiplier
   FM_i = Lifting frequency multiplier associated with the frequency of subtask i
   FM_{i+j} = Lifting frequency multiplier associated with the sum of frequency of subtask i and frequency of subtask j

2. Frequency independent lifting index (FILI) computation

   $FILI$ = Load weight $(L)/(C \times VM \times DM \times HM \times AM \times CM)$
   $FILI_{2 \text{ to } n}$ = FILI from the worst (second subtask) to the best (nth subtask)

3. Composite lifting index (CLI) computation

$$\begin{aligned} \text{CLI} = \text{LI1} &+ \\ &\text{FILI}_2 \times (1/\text{FM}_{1+2} - 1/\text{FM}_1) + \\ &\text{FILI}_3 \times (1/\text{FM}_{1+2+3} - 1/\text{FM}_{1+2}) + \\ &\text{FILI}_4 \times (1/\text{FM}_{1+2+3+4} - 1/\text{FM}_{1+2+3}) + \\ &\text{FILI}_5 \times (1/\text{FM}_{1+2+3+4+5} - 1/\text{FM}_{1+2+3+4}) \end{aligned}$$

FIGURE 6.3 Final formula for computing the CLI.

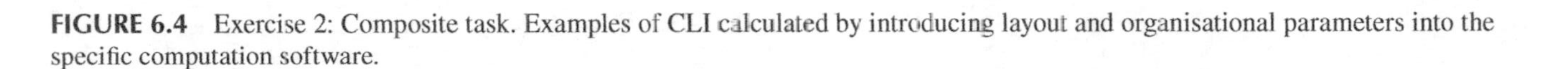

	ACTION DESCRIPTION	Load weight	V: Vertical location (cm)		D: Vertical travel distance (cm)			H: Horizontal location (cm)		A: Asymmetry angle [°]		C: Coupling		F: Freq	Work duration (min)		No. hands		No. workers		RWL man	LI man
1	A	5.0	10	0.81	40	50	0.93	45	0.56	0	1.00	P	0.90	2.40	60	0.90	2	1.00	1	1.00	8.4	0.59
2	B	5.0	50	0.93	0	50	1.00	40	0.63	0	1.00	P	0.90	2.40	60	0.90	2	1.00	1	1.00	11.7	0.43
3	C	5.0	90	0.96	40	50	0.93	40	0.63	0	1.00	P	0.90	2.40	60	0.90	2	1.00	1	1.00	11.2	0.44
4	D	5.0	130	0.84	80	50	0.88	40	0.63	0	1.00	P	0.90	2.40	60	0.90	2	1.00	1	1.00	9.2	0.54
5	E	5.0	170	0.72	120	50	0.86	40	0.63	0	1.00	P	0.90	2.40	60	0.90	2	1.00	1	1.00	7.7	0.65

CLI men (age 18–45)	1.31	RISK PRESENT
CLI men (age <18 or >45)	1.63	RISK PRESENT
CLI women (age 18–45)	1.63	RISK PRESENT
CLI women (age <18 or >45)	2.18	RISK PRESENT

FIGURE 6.4 Exercise 2: Composite task. Examples of CLI calculated by introducing layout and organisational parameters into the specific computation software.

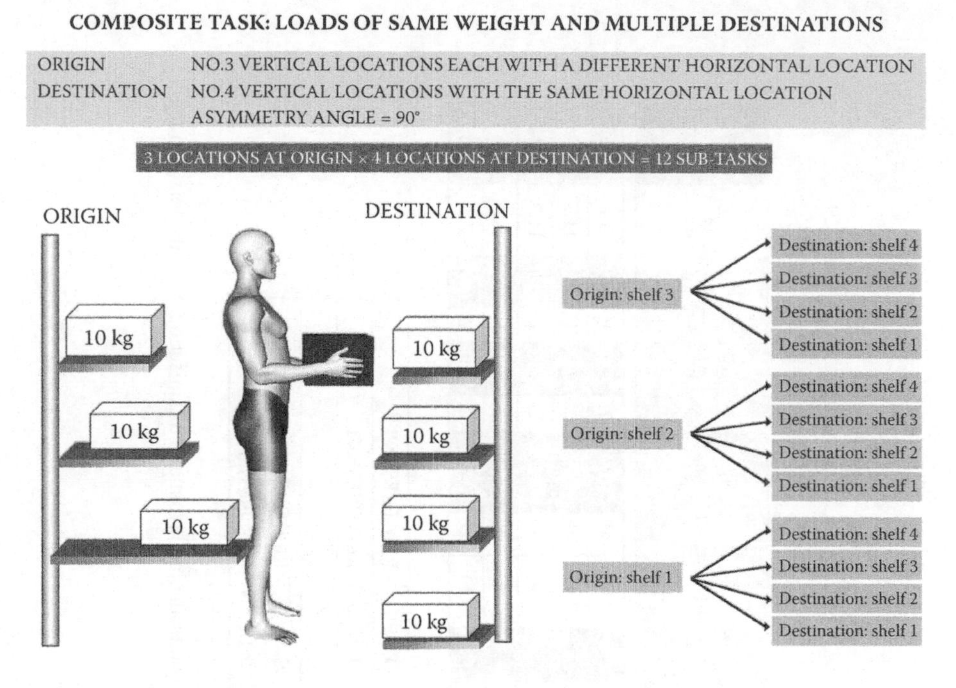

FIGURE 6.5 Exercise 3: Composite task. Determining subtasks.

could be not reliable [Waters et al., 2009]. Dividing overall frequencies by too many subtasks brings the result of individual intrinsic frequencies too close to zero, causing the risk factor deriving from the organisational aspects to be underestimated (the frequency multiplier (FM) would always tend to be equal to 1).

This section illustrates several examples of determining the number of subtasks, while the next chapter goes on to analyse more complex composite tasks featuring a high number of subtasks (more than 10) and the simplification methods required to obtain an accurate analysis of the lifting index LI.

Figure 6.5, Exercise 3, shows a lifting task with three different heights at origin (each at the same horizontal distance from the body) and four different heights at destination (also at the same horizontal distance from the body); in this example there are 12 subtasks.

Figure 6.6, Exercise 4, features another lifting task with six heights at origin, three distances from the body at origin, three heights at destination, and two horizontal distances at destination.

This lifting task, where only one kind of object with a constant weight is considered, has a layout configuration that determines the existence of 108 potential subtasks to be analysed (6 × 3 × 3 × 2 = 108).

In these two examples, particularly the second one, it is no longer possible to accurately compute the CLI using the classic criteria: The number of subtasks needs to be reduced using methods that will be explained in the next chapter.

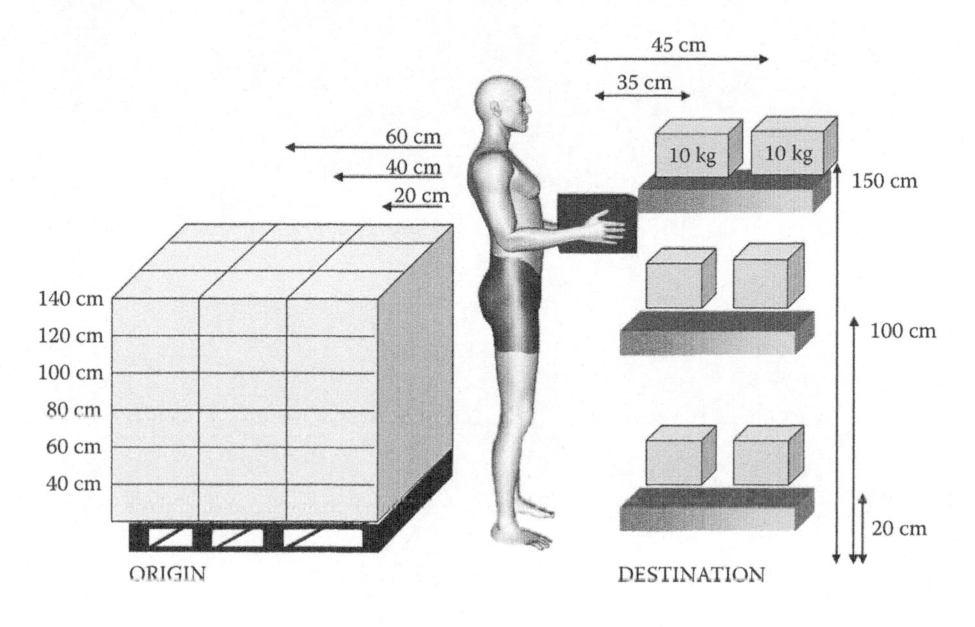

FIGURE 6.6 Example 4: Composite task. Determining subtasks.

7 Computing the CLI Exposure Index for Evaluating Complex Composite Tasks (with More Than 10 Subtasks): Procedures and Examples

7.1 SIMPLIFICATION CRITERIA AND PROCEDURE FOR LAYOUT VARIABLES

It has been emphasised that while it may be easy to compute a lifting index for *single task* handling situations, it is much more difficult to compute the index for lifting tasks that involve lifting and lowering objects to and from shelves with different heights and depths (reaches). Computing *composite tasks* may require many more variables and thus require more time to analyse, with the problem of generating evaluation errors.

Recalling the formula for computing composite tasks lifting index (CLI), the main problem is that an excessive number of subtasks could make the frequency multiplier irrelevant: in other words, an increased number of subtasks may lead to underestimating risk.

To overcome this problem, the best approach is to streamline the subtasks: By reducing them, the frequency will be divided by a smaller number, thus solving the problem at the outset.

To accurately calculate the CLI when there are too many subtasks in a composite lifting task (for example, more than 10), it is necessary to reduce the number of subtasks. It is only by adopting this simplification procedure that the original equation for CLI [Waters et al., 1994] can be properly applied [Waters et al., 2009] [Colombini et al., 2009].

Failure to precisely define the simplification criteria may lead to evaluation discrepancies or errors. Therefore, since work environments often feature composite tasks (and variable tasks, to be explained in Chapter 8), a working group was formed that, with Thomas Waters, the main author of the original method, defined the simplification criteria for both the computation variables and methods.

The final recommendations are presented here, and should be adopted for accurate simplifications, considering the different variables included in the equation.

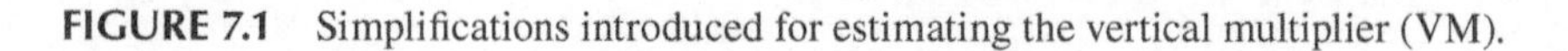

	CM	AVERAGE VALUE (CM)	MULTIPLIERS
IDEAL AREA	51–125	75	**1**
NON IDEAL AREA	0–50 OR 126–175	0 OR 150	**0.78**
CRITICAL AREA	ABOVE 175 CM OR BELOW 0 CM		

FIGURE 7.1 Simplifications introduced for estimating the vertical multiplier (VM).

7.1.1 Vertical Multiplier (the Distance of the Hands from the Floor at the Origin/Destination of Lifting)

The vertical multiplier (i.e., the height of the hands at the origin/destination of lifting) has been reduced in two different areas (Figure 7.1):

- *Ideal area*: The hands are between 51 and 125 cm; the resulting vertical multiplier (VM) is defined as equal to 1.
- *Nonideal area*: The hands are below 50 cm or above 125 cm. The resulting VM is defined as equal to 0.78.

The *critical area* above 175 cm (or below 0 cm) is to be regarded as absolutely unacceptable (multiplier = 0.00) and should be avoided (critical code).

7.1.2 Horizontal Multiplier (Maximum Distance between the Load and the Body during Lifting)

The horizontal distance has been simplified in three different areas (Figure 7.2):

- *Ideal area (near)*: The horizontal location is equal to or less than 40 cm; the resulting horizontal multiplier (HM) is defined as 0.71 (for a representative value of 35 cm).
- *Nonideal area (medium)*: The horizontal location is between 41 and 50 cm; the resulting HM is defined as 0.56 (for a representative value of 45 cm).
- *Nonideal area (far):* The horizontal location is between 51 and 63 cm; the resulting HM is defined as 0.40 (for a representative value of 63 cm).

The *critical area* over 63 cm is to be regarded as absolutely unacceptable and should be avoided (critical code).

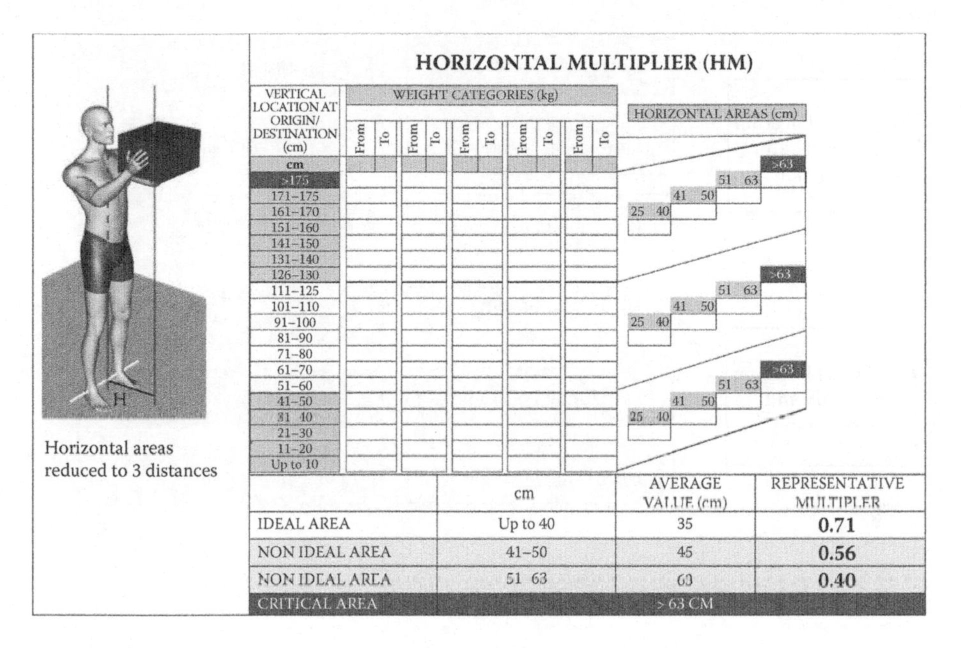

	cm	AVERAGE VALUE (cm)	REPRESENTATIVE MULTIPLER
IDEAL AREA	Up to 40	35	**0.71**
NON IDEAL AREA	41–50	45	**0.56**
NON IDEAL AREA	51–63	63	**0.40**
CRITICAL AREA		>63 CM	

FIGURE 7.2 Simplifications introduced for estimating the horizontal multiplier (HM).

7.1.3 Asymmetric Multiplier (the Angular Measure of Displacement of the Load)

Body twists are briefly considered for all lifting actions in a composite task (for variable tasks—see Chapter 8—where objects of different weights are lifted, they are considered for each weight category). The asymmetric multiplier (AM) (value 0.81) is assigned to all subtasks if the angle of asymmetry exceeds 45° for at least 50% of the lifting actions. Alternatively, the asymmetric multiplier is equal to 1.

7.1.4 Vertical Travel Distance Multiplier (Vertical Travel Distance of the Hands between the Origin and the Destination of the Lift)

The contribution of this factor has been considered as noninfluent. The corresponding multiplier (DM) has thus been taken as a constant, equal to 1. It should be stressed that even if the vertical distance multiplier (DM) is set as a constant, the height of the hands at both the origin and the destination of the lift should be considered and measured.

7.1.5 Coupling Multiplier

The contribution of this factor has also been defined as constant. Experience has taught that ideal couplings are very rare, so the corresponding multiplier (CM) is defined as a constant equal to 0.90.

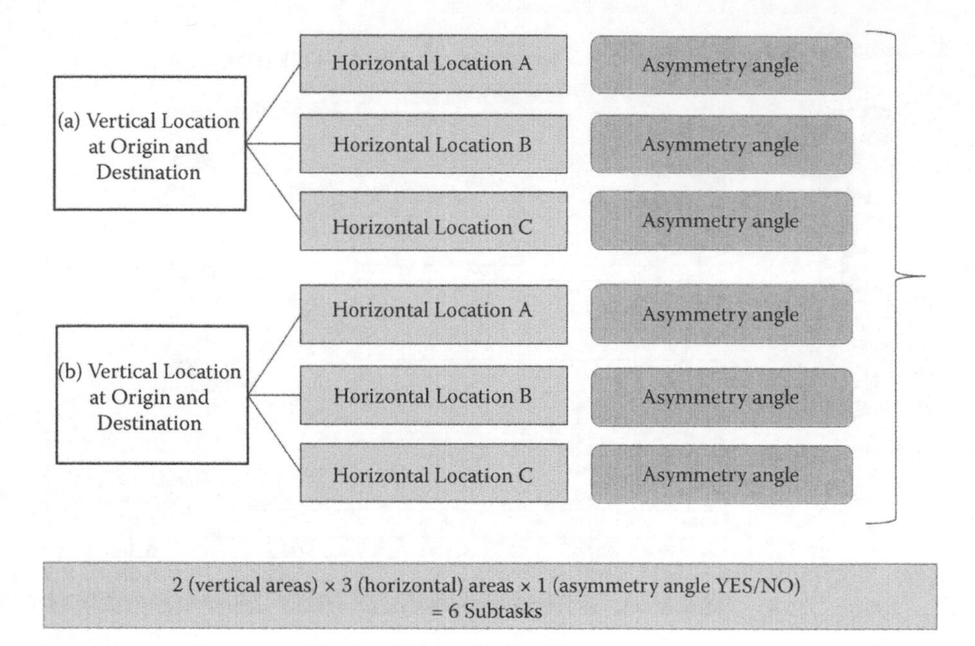

FIGURE 7.3 Grouping results into six subtasks.

7.1.6 Grouping the Results and Computing the Final CLI

By adopting the procedure illustrated here, it is therefore possible to analyse a scenario defined as a composite lifting task as being characterised by just six subtasks; this allows the traditional equation for computing the CLI to be correctly used.

In fact, the approach produces up to six lifting indexes (LIs) for up to six different subtasks (Figure 7.3):

- Two vertical location areas
- Three horizontal location areas
- One asymmetry condition

For each of these subtasks as well as for the composite task, the specific frequency and duration of lifting are evaluated as described in Chapter 5.

7.2 An Example of a Composite Task Comprised of Over 10 Subtasks Computed Using a Dedicated Software

As an example, the CLI is computed for Exercise 4 (Figure 7.4), which was described in Chapter 6. The task was found to be made up of 108 subtasks ($6 \times 3 \times 3 \times 2 = 108$):

- Six vertical locations at origin
- Three horizontal locations at origin

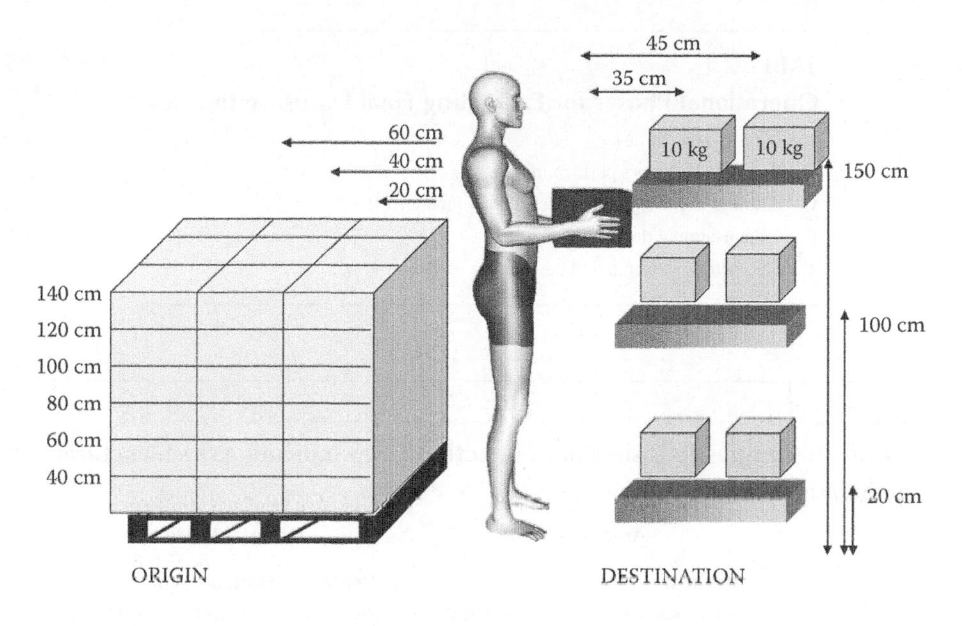

FIGURE 7.4 Exercise 4: Composite task. Description of layout at origin and destination.

- Three vertical locations at destination
- Two horizontal locations at destination

Since there are more than 10 subtasks, it is not possible to accurately compute the CLI using the classic criteria: The number of subtasks needs to be reduced using the methods explained above.

The composite task is now analysed in terms of operational phases.

Table 7.1 provides an overview of the various phases that are followed to obtain a final estimate of the exposure level.

7.2.1 Identifying the Type of Task

Table 7.2 provides all the structural information about the task: It is composite because the object lifted has only one weight but it is lifted with several geometries at the origin and at the destination.

7.2.2 Description of Operators Performing the Task

There is just one operator, who is a healthy adult male. The analysis is performed using dedicated software: ERGOepm_CLI&VLI_S (which can be downloaded free of charge from www.epmresearch.org, in English and in Italian, and from www.epminternationalschool.org in Spanish).

On sheet "1.Key enter" of the software (*quick evaluation*), after filling out these sections and obtaining a result indicating that a full evaluation is required, enter the number of workers performing the task that involves lifting a certain number of objects per shift into the space provided (Figure 7.5).

TABLE 7.1
Operational Phases for Estimating Final Exposure Indexes

A. Identify type of task
B. Describe operators performing the task
C. Number and weight of objects lifted during shift
D. Organisational data: Shift diary
E. Geometries at origin and destination

TABLE 7.2
Exercise 4: Composite Task: Data Collection Form Indicating the Structural Characteristics of the Task

Loads Characteristics		Areas at Origin			Areas at Destination		
Number	Weight	Vertical Locations (cm)	Horizontal Locations (cm)	Asymmetric Angle	Vertical Locations (cm)	Horizontal Locations (cm)	Asymmetric Angle
1,000	8 kg	40	25	Absent	20	35	Absent
		60	41		100	45	
		80	60		150		
		100					
		120					
		140					

Number of workers performing the same task (manual handling task): The whole homogeneous group	1

FIGURE 7.5 Exercise 4: Composite task. Number of workers performing the same task (software: sheet "1.Key enter").

7.2.3 Weight and Number of Objects Lifted during the Shift

On sheet "2.Production data" of the software, enter into column (a) how many units the individual worker lifts (if there is only one worker) or how many units the whole group lifts, per shift (Figure 7.6).

Enter into column (b) how often units belonging to the same category are lifted: if lifted just once, enter 1.

The software automatically calculates the number of lifts per shift (and consequently the overall number of lifting during the shift).

If objects belonging to the same weight class are almost always lifted simultaneously by more than one worker (sheet 1), enter the number of workers in the specific

	Weight (kg)	Number of Objects Lifted per Shift by the Whole Group (*a*)	Number of Lifts per Unit of Weight (*b*)	Total Number of Objects Lifted per Shift by the Whole Group (*a* × *b*)
From 3 to 3.99	3.5			0
From 4 to 4.99	4.5			0
From 5 to 5.99	5.5			0
From 6 to 6.99	6.5			0
From 7 to 7.99	7.5			0
From 8 to 8.99	8.5	1,000	1	1,000
From 9 to 9.99	9.5			0
	Total			1,000

FIGURE 7.6 Exercise 4: Composite task. Weight and number of objects lifted during the shift (software: sheet "2.Production data").

Number of workers simultaneously lifting this weight category (0.85 multiplier)	Weight category lifted by 1 upper limb: Mark *X* (0.6 multiplier)

FIGURE 7.7 Exercise 4: Composite task. Exposure index adjustments for lifts performed by two or more workers or with one hand (software: sheet "2.Production data").

box shown in Figure 7.7. The software will automatically adjust the average weight for the category and the final computation of the risk index will take the event into account (see details on person multiplier in Chapter 4).

If the majority of the lifts (for one weight category) are performed using one hand, enter an X in the appropriate box of the software (Figure 7.7).

The software will automatically adjust the final computation of the index to take this event into account (see details on one-hand multiplier in Chapter 4).

7.2.4 Collecting Organisational Data: The Shift Diary

The shift diary (Figure 7.8) records lifting and nonlifting operations or activities relating to the job. It is of fundamental importance for calculating the frequency and duration scenario of lifting tasks.

Example 4 refers to a single 8 h shift (480 min) with 4 h in the morning, a 1 h lunch break (not counted as working hours, as usual in daily single shifts), and 4 h in the afternoon. There are periods of manual lifting alternating with periods of light work or breaks, and jobs that involve manual pulling/pushing.

The organisational data, to be entered into the software (sheet 3) in the order of occurrence (leave the space blank if the event does not occur), includes:

	Other tasks or breaks	MANUAL LIFTING TASK (including carrying)	Other tasks or breaks	Pushing pulling task	Other tasks or breaks	MANUAL LIFTING TASK (including carrying)	Other tasks or breaks	Pushing pulling task	Other tasks or breaks	MANUAL LIFTING TASK (including carrying)	Other tasks or breaks	Pushing pulling task	Other tasks or breaks	MANUAL LIFTING TASK (including carrying)	Other tasks or breaks	Pushing pulling task	Other tasks or breaks	MANUAL LIFTING TASK (including carrying)	Other tasks or breaks	Pushing pulling task	Other tasks or breaks
Minutes	60	**60**		30		**60**		30	60	**60**		30		**60**		30		**30**		15	15
START	**8.00**																				END
Note									LUNCH												
Hours in the shift	9.00	10.00		10.30		11.30		12.00	13.00	14.00		14.30		15.30		16.00		16.30		16.45	17.00
Push-pull (min)				30				30				30				30				15	

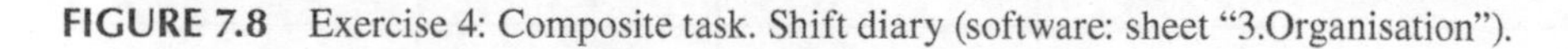

FIGURE 7.8 Exercise 4: Composite task. Shift diary (software: sheet "3.Organisation").

- *Light work and breaks* (periods of no less than 10 min).
- *Manual load lifting* (periods of no less than 30 min) *including manual carrying times* (objects weigh 3 kg or more).
- *Manual pushing/pulling* (periods of no less than 10 min).
- Shift start time: The start times of all the operational phases described. (The shift finishing time will appear automatically when all the organisational input is completed.)
- Add any remarks, such as the duration of the meal break, in the space marked notes.

For shifts that alternate lifting periods of less than 30 min and light work of less than 10 min, add the period of light work to the lifting period: The lifting frequency will diminish as a consequence.

It is also useful to deduct any net time devoted to the manual carrying of objects proper from the total manual lifting and carrying times.

The percent of carrying loads should be indicated in the appropriate boxes in sheet 2.

Once the shift diary is completed, many organisational details will be shown on sheet "3.Organisation" (Figure 7.9), such as:

- Shift duration in minutes. It is important to remember that if the meal break is not included in the working hours (as in this example) its duration must be entered into the appropriate box for this session.
- Total amount of time (minutes) devoted to manual lifting and carrying.
- Net duration of pushing/pulling activities. This parameter is not used for computing the lifting frequency but is extremely useful for indicating the need to complete a risk analysis also for these two load handling activities.
- Number of lifts performed by the whole group (if two or more workers are involved).
- Number of lifts performed by each individual worker.
- Overall lifting frequency (per worker) as number of lifts per minute.

Long Duration	
Number of workers involved	1
Duration of meal break in minutes (indicate only if not included in the work shift)	60
Shift duration (min)	480
Net duration of MMH in a shift including carrying (min)	**270**
Net duration of pushing and pulling (min)	135
Total number of objects lifted (weight more than 3 kg) by the group	1,000
Total number of objects lifted (weight more than 3 kg) by each worker	1,000
Lifting frequency (lifts/min)	**3.70**

FIGURE 7.9 Exercise 4: Composite task. Computation of frequency and duration (software: sheet "3.Organisation").

7.2.5 Analysis of Geometries at Origin and Destination

It is relatively easy to enter the data for the position of objects at origin and destination (Figure 7.10).

Sheet "4.Workplace description" of the software includes 2 boxes that simulate the shape of benches or shelves in a three-dimensional view: One box is for entering the position of the loads at the origin and the other for position at the destination.

Place an *X* in the two boxes to indicate the areas (height from the floor or distance from the body) from which the objects are lifted at the origin (upper box) and where they are placed at the destination (lower box). These areas should be specified in the column indicating the relevant weight category (just one for composite tasks).

If tasks require twisting of the upper body, place an *X* only when the twist is more than 45° and is repeated for more than 50% of the lifts.

In the example there is no upper body twisting.

7.2.6 Generating a Table of Aggregated Data

Using the data collected up to now it is possible to generate a table of data for six subtasks similar to the one already presented in Chapter 6 for computing traditional CLI. Table 7.3 reports data for Exercise 4. The software calculates this part without showing it.

7.2.7 Interpreting the Final Results

It is now possible to compute the final CLI by using its traditional formula (see Chapter 6). Using the software, the final results indicated in the sheet "5.LI" (lifting index) can now be read.

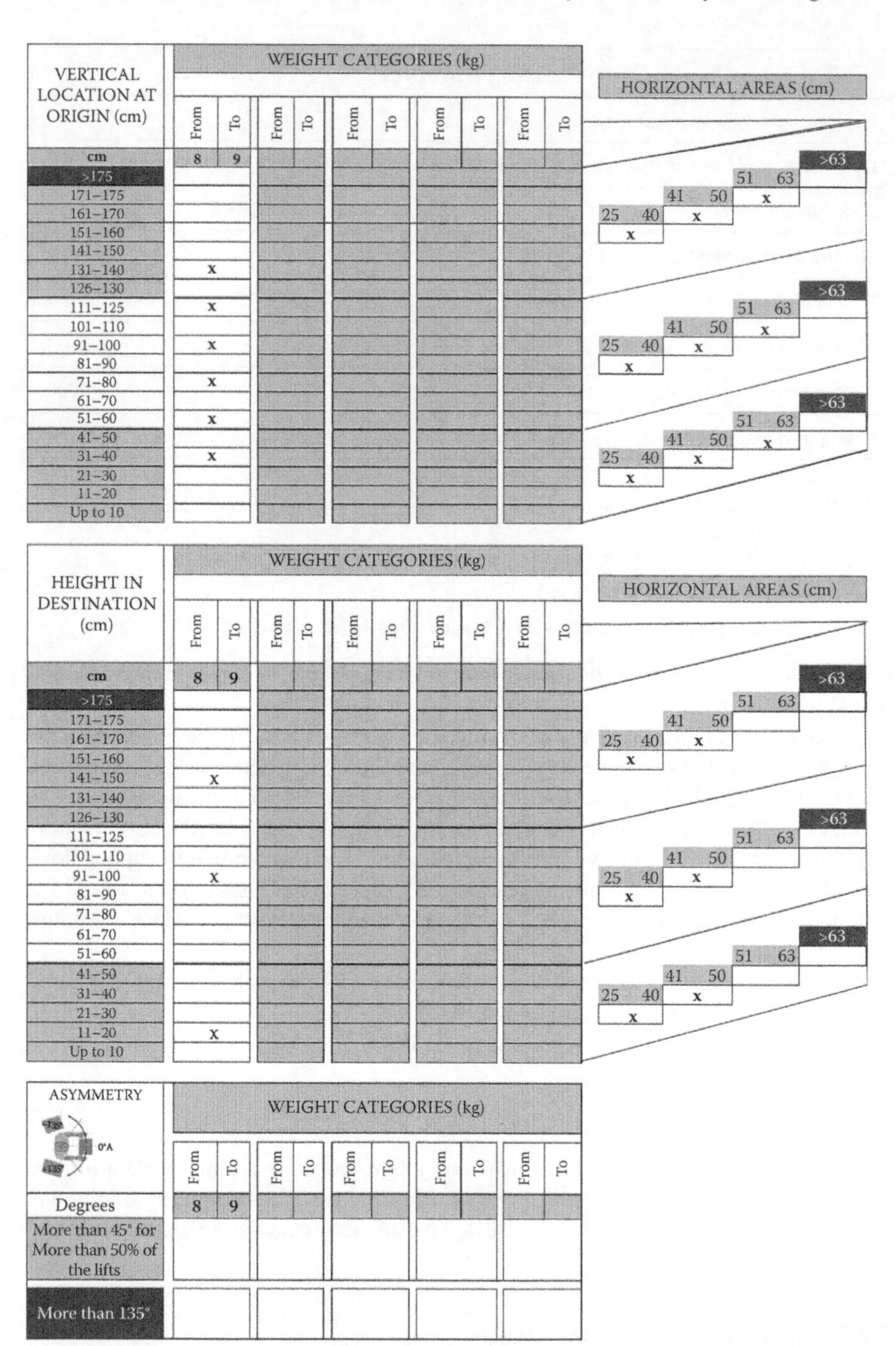

FIGURE 7.10 Exercise 4: Composite task. Entry of data describing load geometries (load position) at origin and destination and description of asymmetry (software: sheet "4.Workplace description").

TABLE 7.3

Exercise 4: Composite Task: Relevant Data for Computing CLI after Aggregation into Six Subtasks

Subtask	Weight (kg)	Vertical Height Classification and VM		Vertical Dislocation (DM = Constant = 1)		Horizontal Distance and HM		Asymmetry		Coupling (CM = Constant = 0.9)		FIRWL	FILI	Frequency (Rounded)	Duration Scenario and FM (Rounded)		RWL	STLI
1	8.5	*L/H*	0.78	*G*	1	*N*	0.71	*A*	1	*P*	0.9	12.46	0.68	0.66	*LD*	0.79	9.85	0.86
2	8.5	*L/H*	0.78	*G*	1	*M*	0.56	*A*	1	*P*	0.9	9.83	0.86	0.66	*LD*	0.79	7.77	1.09
3	8.5	*L/H*	0.78	*G*	1	*F*	0.40	*A*	1	*P*	0.9	7.02	1.21	0.33	*LD*	0.83	5.85	1.45
4	8.5	*G*	1	*G*	1	*N*	0.71	*A*	1	*P*	0.9	15.98	0.53	0.82	*LD*	0.77	12.33	0.69
5	8.5	*G*	1	*G*	1	*M*	0.56	*A*	1	*P*	0.9	12.60	0.67	0.82	*LD*	0.77	9.72	0.87
6	8.5	*G*	1	*G*	1	*F*	0.40	*A*	1	*P*	0.9	9.00	0.94	0.41	*LD*	0.82	7.40	1.15

European Standard: EN 1005-2, ISO Standard: 11228-1				
kg	**Protected Population**	**Lifting Index**		
25	Men (18–45 years old)	2.05		Risk present
20	Women (18–45 years old)	2.57		Risk present
20	Men (<18 or >45 years old)	2.57		Risk present
15	Women (<18 or >45 years old)	3.42		Risk present
Revised NIOSH Lifting Equation				
23	NIOSH original	2.23		Risk present

FIGURE 7.11 Exercise 4: Composite task. Final CLIs per gender and age groups (software: sheet "5.LI").

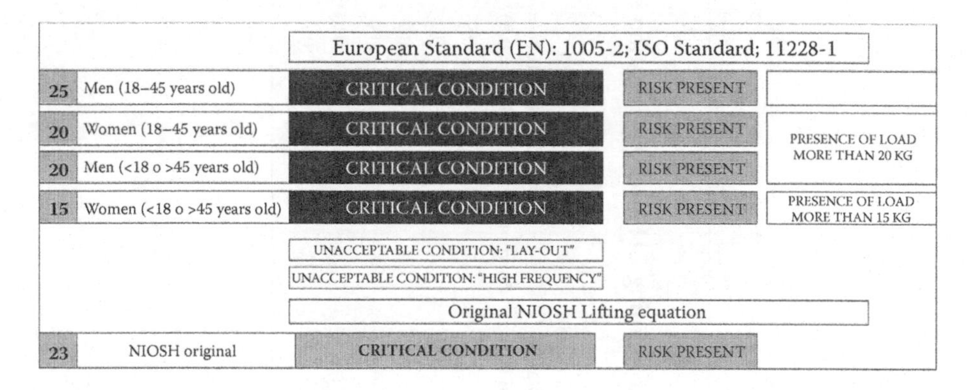

FIGURE 7.12 General example of critical conditions: Critical codes (software: sheet "5.LI").

As shown in Figure 7.11, the results will differ based on the gender and age group of the workers. The CLI is indicated also using the original revised NIOSH lifting equation (RNLE) formula that considers a weight constant of 23 kg (regardless of gender or age).

If conditions defined as critical are detected and entered into the program (i.e., geometries to be avoided, excessive weights or frequencies—see Chapter 4), the software will highlight them in the final computation of the exposure indexes, with explanations (Figure 7.12).

7.3 A MATHEMATICAL DEMONSTRATION: COMPUTATION ERROR IN THE EVENT OF EXCESSIVE BREAKDOWN OF FREQUENCY

As mentioned earlier, if there are too many subtasks, this may cause one or more of the addenda in the formula to be annulled due to excessively breaking down the frequency and thus the relevant multiplier.

This is the formula for calculating the CLI:

$$CLI = LI_1 + \sum_{i=2}^{n} \left\{ FILI_i x \left(\frac{1}{FM_{1+...+i}} - \frac{1}{FM_{1+...+(i-1)}} \right) \right\} \tag{7.1}$$

where the numbering going from 1 to n is derived from the decreasing order by risk level of the various LI_i.

Or, more simply, the equation is developed for only three subtasks:

$$CLI = LI_1 + FILI_2 x \left(\frac{1}{FM_{1+2}} - \frac{1}{FM_1} \right) + FILI_3 x \left(\frac{1}{FM_{1+2+3}} - \frac{1}{FM_{1+2}} \right) + [...] \tag{7.2}$$

The FMs in the equation are the frequency multipliers calculated as a function of the sum of the frequencies of the subtasks indicated with subscripts: FM_{1+2} is the frequency multiplier produced by the sum of the frequencies of subtasks 1 and 2, FM_{1+2+3} is the frequency multiplier produced by the sum of the frequencies of subtasks 1, 2, and 3, and so on.

Let us now suppose that the frequency contributed by the third subtask is minimal; this will result in

$$FM_{1+2} = FM_{1+2+3} \tag{7.3}$$

Under these circumstances, the multiplier enclosed by brackets, relative to $FILI_3$, will cancel out and consequently eliminate the entire contribution of the third subtask to the total risk:

$$CLI = LI_1 + FILI_2 x \left(\frac{1}{FM_{1+2}} - \frac{1}{FM_1} \right) + FILI_3 x \left(\frac{1}{FM_{1+2+3}} - \frac{1}{FM_{1+2}} \right) + [...]$$

Obviously the higher the number of subtasks at sufficiently low frequencies, the greater the effect of this event.

Therefore the event translates into the loss of the contribution of the cancelled subtasks to the overall risk: The lifting index would be influenced by the error introduced by eliminating one or more tasks.

By introducing standard simplifications, the methodology described here prevents this event from occurring, which would otherwise lead to underestimating the overall risk.

7.4 FINAL COMMENTS FOR EVALUATING CLI USING DIFFERENT MODELS OF SOFTWARE PROPOSED BY THE AUTHORS

When the subtasks' number is equal to or less than 10, the CLI will have more precise results using the classic CLI evaluation proposed by NIOSH [Waters et al., 1994]. With the software ERGOepm_LI&CLI_C, it is possible to obtain automatically the lifting frequency and the corresponding FM, describing accurately the group of workers involved in manual lifting (sheet 1), the number of lifted loads (sheet 2), and the shift diary (sheet 3). For the layout description the scheme is the same as that proposed in Figure 6.4: The total frequency has to be split equally between the present subtasks.

On the contrary, when the subtasks' number is more than 10, the new methodology proposed in this chapter and the ERGOepm_CLI&VLI_S software have to be used.

8 Variable Lifting Tasks—The Variable Lifting Index (VLI): Computation Procedures and Examples

8.1 GENERAL ASPECTS AND PROCEDURES

The term *variable task* is used to describe a job that involves lifting and lowering different objects with different weights over different geometries (e.g., horizontal reach, vertical height, etc.).

The notion of a variable lifting index (VLI), a new method for estimating the level of physical stress associated with lifting variable loads, was first presented at the International Ergonomics Association conference in Beijing by the authors with two different presentations [Waters et al., 2009] [Colombini et al., 2009].

The concept for the VLI assessment method is similar to the composite lifting index (CLI) method for composite task jobs. The difference is that rather than using individual task elements, all of the lifts will be first aggregated into a maximum of 30 subtasks and correspondent frequency independent lifting index (FILI), which in turn will be distributed into a fixed number of FILI categories (e.g., six), each with a variable frequency. These six FILI categories will then be weighted using the traditional CLI equation.

The VLI approach asks for a systematic organisational analysis and is based on a comprehensive assessment of the lifting tasks in order to analytically determine:

- The overall duration of the variable lifting task in the shift.
- The number of objects of different weight lifted during this time.
- The number of workers involved, thus deriving (for one representative worker) the overall frequency of lifts (and the corresponding duration scenario).
- The partial frequency of lifts for each weight (or group of similar weights).
- The approximate frequencies of individual lifts (subpartial frequencies) according to different geometries of lifts for each group of similar weights. These are derived by direct observation of workplaces and using a probabilistic computation.

This approach is mainly based on production or sales data (for lifting durations, number of loads, lifting frequencies, etc.) or probability distribution data (for geometries and subpartial frequencies). In order to use this approach several simplifications may be necessary, especially when many and different objects are lifted along several and casual geometries: This is because otherwise the number of individual subtasks and corresponding FILI values could add up to be very high (hundreds and sometimes thousands) and be practically impossible to manage. Simplifications regard both grouping the weights lifted during a shift (as will be discussed later), both criteria for analysing geometries of lifts, and determining the different multipliers to be used in the revised NIOSH lifting equation (RNLE) (as already presented in Chapter 7).

The number of potential individual lifting tasks in the job has to be aggregated into a structure that considers up to a maximum of 30 subtasks and the corresponding frequency independent lifting index (FILI) and single task lifting index (STLI) for different loads (weight categories) and geometries using the following:

- Aggregate up to five objects (weights) categories.
- Classification of vertical location (VM) in only two categories (good/bad).
- Classification of horizontal location (HM) in up to three categories (near, medium, far).
- Indicating the presence/absence of asymmetry (AM) assessed for each weight category (by threshold value for all the lifts in the category).
- Determining the daily lifting duration classified as in the applications manual [Waters et al., 1994].
- Determining the frequencies of lifts specifically determined or estimated for each subtask and FILI; frequency multipliers (FMs) are determined as in the applications manual [Waters et al., 1994]. Estimation of the frequencies for each resulting subtask could be performed by two different approaches (or models) that will be illustrated in this chapter.
- Vertical displacement (DM) and coupling (CM) are both considered constants.
- At the end it will be possible to compute individual FILI and STLI for up to 30 subtasks.

Once the selected FILI values are derived, the entire set of FILI values are assigned into a fixed number of LI categories. We suggest six categories be used. The categories can be defined by assigning the FILI values according to "sextiles" of the correspondent FILI distribution. This primary choice allows having an equal (or similar) number of FILI values in each of the six categories. However, since it is not easy to identify manually the key values corresponding to the sextiles of the given distribution, a simpler alternative for manually grouping LI categories involves dividing the range of FILI values (from minimum to maximum) into six and consequently assigning all the resulting FILI values: This alternative has the minor disadvantage that some LI categories could be empty.

Once the selected FILI values are assigned to the appropriate category, the corresponding cumulative frequencies can be computed and applied using the traditional CLI equation with frequency weighting. Although the calculations can be performed by hand, dedicated software is most helpful in performing these difficult computations. Two software applications were ad hoc prepared: ERGOepm_CLI&VLI_S, the more synthetic model, and ERGOepm_VLI_AP, the more analytic model. We will present examples of how to apply the VLI using the software in the next pages.

8.2 SIMPLIFICATION CRITERIA AND PROCEDURES FOR LOAD AND LAYOUT VARIABLES

8.2.1 Collecting and Entering Organisational Data into the Software

The first step in the analysis is to identify the workers and how many workers (one or more) are involved in similar load lifting activities (see in the above-mentioned software sheet "1.Key enters").

The data to be collected include the weight of the object (from 3 kg to the maximum weight lifted, divided into 1 kg increments) and the relative number of objects lifted in a shift by one worker (if there is only one) or by the selected group of workers (see sheet "2.Production data").

Production (or sales) data are generally known to management: In some countries, there is a legal requirement to know the weight of the object to be lifted, which must be indicated on the packaging.

It should be noted that if the same loads have to be lifted systematically (or lowered) from origin to destination two or more times, the number of lifted loads, to report as lifted, will increase. If there are more than five types of weights, the weights will be aggregated (automatically by the software) into a maximum of five weight categories, on a statistical basis, varying according to respective type and quantity. A representative average (by frequency) weight is selected for each category.

Table 8.1 shows an example of this aggregation process. The range (max value – min value) of lifted weights (A) is divided in five equal intervals for identifying weight categories (B). The weights lifted with correspondent number are aggregated into those categories. An average (by frequency) weight is computed for each weight category and will represent it in consequent computations.

From organisational data (see also Chapter 7) such as "number of workers involved in the task(s)," "net duration of manual lifting in the shift," "total number of objects lifted during a shift," and "number of objects within each weight category lifted during a shift," it is possible to determine the overall lifting frequency (per worker) and the partial lifting frequency for each weight category (as shown at part B in Table 8.1) and to use the analytical corresponding frequency multipliers (FMs) from traditional tables [Waters et al., 1994], considering the appropriate lifting duration scenario (short, medium, long).

The proposed software will help in computing all data regarding weights lifted into up to five weight categories. Small weight categories (expressed in kg) are preentered into the software, so that the loads falling within each 1 kg range (e.g., 4.2, 4.3, 4.8, 4.9 kg) are counted and grouped together (e.g., from 4 to 4.99 kg). The computation then uses the average weight of the category (e.g., 4.5 kg).

For each small (1 kg) category, the number of objects lifted manually per shift either by the individual worker (if there is only one) or by the entire group has to be indicated.

If the object is lifted and lowered more than once, the number of times it is handled is also indicated, and this will change the number of loads actually lifted.

The indicated loads are (automatically by the software) redistributed into five weight categories, whose type and number may differ from case to case to more accurately represent the loads actually lifted. This is because the categories are computed

TABLE 8.1
Aggregation of Several Weights Lifted (by a Worker) during a Shift in Five Weight (WT) Categories and Computation of Correspondent Lifting Frequency: Example for Weights Ranging from 5 to 15 kg and for a Lifting Duration of 300 min in a Shift

A

Weight (WT) (kg)	Number of Objects Lifted per Shift by a Worker
5	100
6	80
7	70
8	100
9	50
10	60
11	40
12	50
13	15
14	10
15	5
Total	580

B

Resulting WT Categories			Number of Objects in Category	Average WT (for the category)	Percentage (%) Objects in Category	Overall and Partial (by WT Category) Frequencies (lpm) (referred to 300 min lifting duration)
Category Number	From	To				
1	5.0	6.0	180	5.4	31.0%	0.60
2	7.0	8.0	170	7.6	29.3%	0.57
3	9.0	10.0	110	9.5	19.0%	0.36
4	11.0	12.0	90	11.6	15.5%	0.30
5	13.0	15.0	30	13.7	5.2%	0.10
						Overall 1.93

Min. WT Value (*a*)	Max. WT Value (*b*)	Difference ($c = b - a$)	ΔWT (for 5 categories)
5	15	10	2

based on the minimum and maximum weights being analysed, which determine the maximum variability range: In other words, all the objects lifted will fall within this range. Later, this range will be divided into five sublevels, thus generating the weight categories. Depending on the minimum and maximum levels of each weight category, the various loads lifted will fall within a specific category. Based on this distribution it is possible to reconstruct:

- The number of loads belonging to each weight category
- Their prevalence in percentage terms over the total number of objects lifted (and if referred to total duration of lifting task, the partial frequency of lifts per minute for each weight category)
- The average weight of each category with respect to the total number of objects belonging to the same category

It is worth noting that based on the intrinsic variability of the categories it is possible for one (or more) category to be blank. This may happen when the weights of the objects lifted are not symmetrically distributed as in the example of Table 8.2.

When this happens, the blank category (Table 8.3) contributes nothing to the lifting index.

In the example, the minimum load is 3.5 kg while the maximum load is 14.5 kg. In this case, there are four weight categories. The 10.1 to 12.3 kg category is empty because no loads fall into this range. This latter weight category will have no effect on the final index, which will be based only on the remaining categories (and their respective geometries).

To further clarify the methods adopted, an example is provided showing how to compute a variable lifting task in a supermarket (Exercise 5, Figure 8.1).

TABLE 8.2
Example of Loads Lifted in a Shift, with Weight Asymmetric Distribution

	Weight (kg)	Number of Objects to Lift per Shift by the Whole Group	Number of Lifting for Each Weight Unit	Number of Objects Really Lifted per Shift by the Whole Group
From 3 to 3.99	3.5	100	1	100
From 4 to 4.99	4.5	200	1	200
From 5 to 5.99	5.5	150	1	150
From 6 to 6.99	6.5	100	1	100
From 7 to 7.99	7.5	50	1	50
From 8 to 8.99	8.5	100	1	100
From 9 to 9.99	9.5			0
From 10 to 10.99	10.5			0
From 11 to 11.99	11.5			0
From 12 to 12.99	12.5			0
From 13 to 13.99	13.5			0
From 14 to 14.99	14.5	500	1	500

TABLE 8.3
Example of Distribution of Loads of Table 8.2 in Four Categories

Categories From	Categories To	Number of Objects	Average Weighted Load (kg)	% Lifted Objects
3.5	5.7	450	4.6	38%
5.7	7.9	150	6.8	13%
7.9	10.1	100	8.5	8%
10.1	12.3	0	0	0%
12.3	14.5	500	14.5	42%

EXERCISE 5

The storage staff are 10 and work on a shift of 6 hours. You can not define precisely what is raised by each operator, other than in average terms.
The loading and unloading of packages is done by trolleys and pallets with shelves placed at different heights, ranging from the floor up to 160 cm, both the origin and the destination.
The body distances are 2 at the origin (approximately 35 and 45 cm) and 1 at destination (35 cm).
Not the rare moments of manual transport can be observed (10 minutes every hour distributed within moments of lifting) and pulling and pushing times (10 minutes per hour). They have 20 minute break after the first 3 hours of work. There are no other official breaks but in fact are commonly made 2 more breaks of 5 minutes.

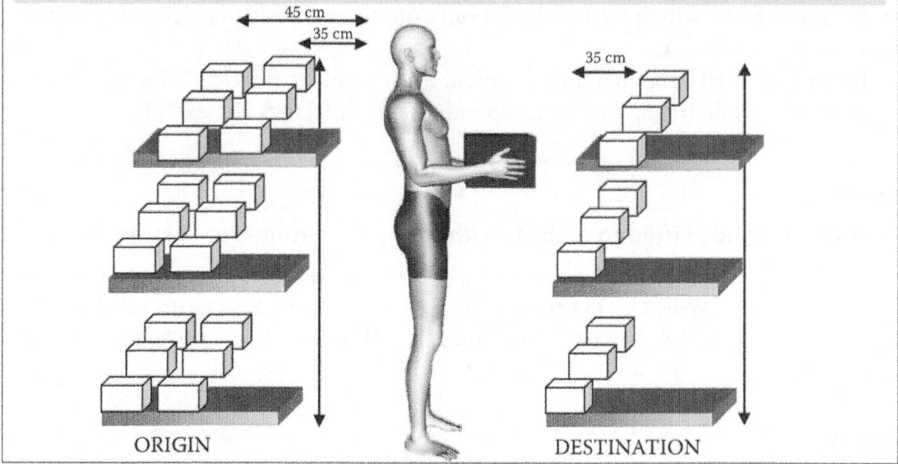

FIGURE 8.1 Exercise 5: Variable task, task description.

Table 8.4 introduces the weight and number of objects lifted as reported in the software at sheet "2.Production data."

The indicated loads are automatically redistributed into five weight categories (Table 8.5).

Begin by entering the number of workers assigned to perform the same task on the first data sheet (sheet "1.Key enters"). In this case a group of 10 workers is assigned to load a certain number of cartons of bottles (wine, water, soft drinks) on to a pallet

TABLE 8.4
Exercise 6: Description of the Geometries at Origin and Destination for the First Weight Category

	Weight (kg)	Number of Objects to Lift per Shift by the Whole Group	Number of Lifts for Each Weight Unit	Number of Objects Really Lifted per Shift by the Whole Group
From 3 to 3.99	3.5	200	1	200
From 4 to 4.99	4.5	790	1	790
From 5 to 5.99	5.5	2,000	1	2,000
From 6 to 6.99	6.5	400	1	400
From 7 to 7.99	7.5	400	1	400
From 8 to 8.99	8.5	1,000	1	1,000
From 9 to 9.99	9.5	800	1	800
Total				5,590

TABLE 8.5
Exercise 5: Variable Task: Distribution of the Lifted Loads in Five Weight Categories (see software sheet "2.Production data")

Categories		Number	Average Weighted	% Lifted
From	To	of Objects	Load (kg)	Objects
4.7	5.9	2,000	5.5	36%
5.9	7.1	400	6.5	7%
7.1	8.3	400	7.5	7%
8.3	9.5	1,800	8.9	32%

during the shift. Since it is hard to know how many cartons each of the 10 workers actually lifts, a statistical average is estimated and used per each worker (it is presumed that the amount of work is equally distributed among the assigned workers).

Table 8.6 describes the shift diary, which specifies the manual lifting and lowering periods of activity, the light work or breaks, and the intervening pulling and pushing activities and correspondent times in minutes (sheet "3.Organisation").

After describing the organisational data, the software will then automatically compute the frequency and duration of the manual lifting task (Table 8.7).

8.2.2 Describing and Evaluating Layout Data

To correctly compute the VLI, certain simplifications regarding geometries need to be introduced, many of which have been described in Chapter 7.

Figure 8.2A illustrates how the positions of the loads (hands) at origin and destination appear in the software ERGOepm_VLI_AP. Each of the five weight categories has a separate space for entering the heights of the loads (sheet "4.Workplace

TABLE 8.6
Exercise 5: Variable Task: Shift Schedule (see software sheet "3.Organisation")

	Other Task or Breaks	Manual Lifting Task (including carrying)	Other Task or Breaks	Push-Pulling Task	Other Task or Breaks	Manual Lifting Task (including carrying)	Other Task or Breaks	Push-Pulling Task	Other Task or Breaks	Manual Lifting Task (including carrying)	Other Task or Breaks	Push-Pulling Task	Other Task or Breaks
Minutes		**60**	**5**	10		**120**	**20**	20		**60**	**5**	10	**50**
Shift starting	**8:00**												End
Shift hours		9:00	9:05	9:15		11:15	11:35	11:55		12:55	13:00	13:10	14:00

TABLE 8.7
Exercise 5: Variable Task: Duration of Manual Lifting and Frequency Evaluation (see software sheet "3.Organisation")

Long Duration	
Number of workers involved	10
Lunch duration in minutes (write only if it is out of the shift duration)	
Shift duration	360
Net duration of MMH in a shift including carrying (min)	**240**
Net duration of and pushing and pulling (min)	40
Total number of objects lifted (weight more than 3 kg) by the group	5,590
Total number of objects lifted (weight more than 3 kg) by each worker	559
Lifting frequency	**2.33**

description"). Vertical heights could be reported considering small intervals of 10 cm each. The horizontal distances of the load from the body (near, medium, far) are then reported, again at both origin and destination, by each weight category, and related to three different groups of vertical distances (bad/low, good/medium, bad/high). This procedure allows detailed classification for each weight category of the possible combinations of vertical height (bad or good) and horizontal distance (near, medium, far).

Since, in practice, it is often very difficult to distinguish all the horizontal distances present in each of the five weight categories, it is (from an applicative point of view) easier and more realistic to describe all together the horizontal areas usually and more representatively reached in the whole variable lifting task (independently from the weight category). Figure 8.2B illustrates how to report, in a simplified manner (ERGOepm_CLI&VLI_S), the positions of the loads (hands) at origin and destination. In this case vertical heights are reported as in the previous scheme, but

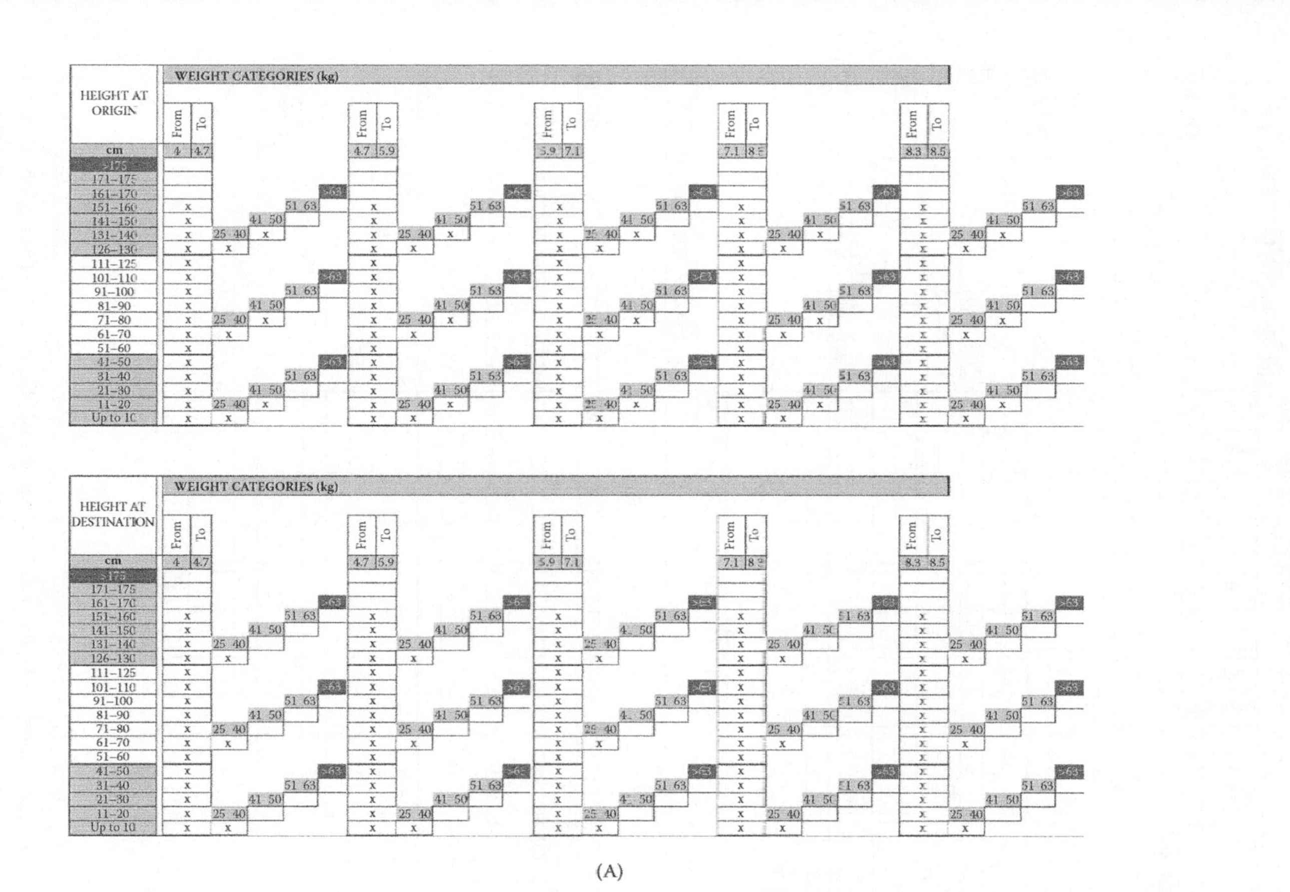

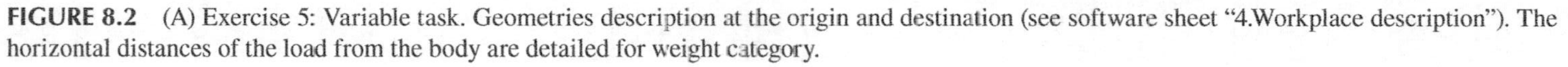

FIGURE 8.2 (A) Exercise 5: Variable task. Geometries description at the origin and destination (see software sheet "4.Workplace description"). The horizontal distances of the load from the body are detailed for weight category.

VERTICAL LOCATION AT ORIGIN (cm)	WEIGHT CATEGORIES (kg) From 3.5 To 4.7	From 4.7 To 5.9	From 5.9 To 7.1	From 7.1 To 8.3	From 8.3 To 9.5
cm					
>175					
171–175					
161–170					
151–160	X	X	X	X	X
141–150	X	X	X	X	X
131–140	X	X	X	X	X
126–130	X	X	X	X	X
111–125	X	X	X	X	X
101–110	X	X	X	X	X
91–100	X	X	X	X	X
81–90	X	X	X	X	X
71–80	X	X	X	X	X
61–70	X	X	X	X	X
51–60	X	X	X	X	X
41–50	X	X	X	X	X
31–40	X	X	X	X	X
21–30	X	X	X	X	X
11–20	X	X	X	X	X
Up to 10	X	X	X	X	X

HORIZONTAL AREAS (cm)	25–40	41–50	51–63	>63
	X	X		
	X	X		
	X	X		

HEIGHT IN DESTINATION (cm)	WEIGHT CATEGORIES (kg) From 3.5 To 4.7	From 4.7 To 5.9	From 5.9 To 7.1	From 7.1 To 8.3	From 8.3 To 9.5
cm					
>175					
171–175					
161–170					
151–160	X	X	X	X	X
141–150	X	X	X	X	X
131–140	X	X	X	X	X
126–130	X	X	X	X	X
111–125	X	X	X	X	X
101–110	X	X	X	X	X
91–100	X	X	X	X	X
81–90	X	X	X	X	X
71–80	X	X	X	X	X
61–70	X	X	X	X	X
51–60	X	X	X	X	X
41–50	X	X	X	X	X
31–40	X	X	X	X	X
21–30	X	X	X	X	X
11–20	X	X	X	X	X
Up to 10	X	X	X	X	X

HORIZONTAL AREAS (cm)	25–40	41–50	51–63	>63
	X			
	X			
	X			

(B)

FIGURE 8.2 ***(Continued)*** (B) Exercise 5: Variable task. Geometries description at the origin and destination (see software sheet "4.Workplace description"). The horizontal distances of the load from the body are not distinguished by weight category.

the horizontal distances of the load from the body (near, medium, far) are not distinguished by weight category, but rather they are assigned (in relation to the three different groups of vertical distances) all the same for all weight categories.

In Exercise 5, Figure 8.2A and B reports how the various objects, irrespective of their weight, can be lifted and positioned from/to any height from 0 to 160 cm at both origin and destination. Moreover, two horizontal locations are described at origin (near, medium) and just one at destination (near), since at destination the loads are first lifted on to the shelf and then later pushed back.

In this example, there is no twisting of the worker's trunk (absence of asymmetry).

As can be derived from Figure 8.2A and B, horizontal distances from the body are always the same at both origin and destination: Consequently, in this case, it is advisable to use the simplified model (Figure 8.2B).

This model (Figure 8.2B) is to be preferred whenever it is not possible to distinguish clearly different horizontal distances (or areas) by weight category: In this case the horizontal distances that are signed have the meaning of describing a modal scenario for all weight categories.

On the contrary, when it is possible to clearly distinguish different horizontal distances (areas) between the weight categories, it will be advisable to use the more analytical model (Figure 8.2A). We will detail the question by further examples in this chapter.

8.2.3 Grouping Lifting Indexes (LIs) and Computing the Final Variable Lifting Index (VLI)

Using a traditional approach, in Exercise 5 there would be an impressive 4,096 subtasks, comprised of:

- Sixteen heights at origin
- Two horizontal distances at origin
- Sixteen heights at destination
- One horizontal distance at destination
- Eight separate weights

$$16 \times 2 \times 16 \times 1 \times 8 = 4{,}096$$

Using the simplified procedure for geometries variables shown in Chapter 7, in addition to the aforementioned grouping of weights into five categories, the resulting scenario in Exercise 5 features 20 different subtasks defined by:

- Five weight categories
- Two vertical heights for the hands: good or central; bad—too low/too high (considering origin and destination together)
- Two horizontal areas (near, medium) at origin and one at destination
- One asymmetry condition (in this case absent)

$$5 \times 2 \times 2 \times 1 = 20$$

Table 8.8 reports analytical data concerning Exercise 5, regarding these 20 subtasks and relative individual FILI and STLI.

TABLE 8.8

Exercise 5: Individual Subtask Characteristics and Derived FILI and LI

Subtask	Weight (kg)	Vertical Height Classification and VM		Vertical Dislocation (*DM* = 1; constant)		Horizontal Distance Classification and HM		Asymmetry		Type of Grasp (*CM* = 0.9; constant)		FIRWL	FILI	Frequency (rounded)	Duration Scenario and FM (rounded)		STLI
1	4.3	*L/H*	0.78	*G*	1.00	*N*	0.71	*A*	1.00	*P*	0.90	12.5	0.35	0.14	*LD*	0.85	0.41
2	4.3	*L/H*	0.78	*G*	1.00	*M*	0.56	*A*	1.00	*P*	0.90	9.8	0.44	0.07	*LD*	1.00	0.44
3	4.3	*G*	1.00	*G*	1.00	*N*	0.71	*A*	1.00	*P*	0.90	16.0	0.27	0.13	*LD*	0.85	0.32
4	4.3	*G*	1.00	*G*	1.00	*M*	0.56	*A*	1.00	*P*	0.90	12.6	0.34	0.07	*LD*	1.00	0.34
5	5.5	*L/H*	0.78	*G*	1.00	*N*	0.71	*A*	1.00	*P*	0.90	12.5	0.44	0.31	*LD*	0.84	0.53
6	5.5	*L/H*	0.78	*G*	1.00	*M*	0.56	*A*	1.00	*P*	0.90	9.8	0.56	0.16	*LD*	0.85	0.66
7	5.5	*G*	1.00	*G*	1.00	*N*	0.71	*A*	1.00	*P*	0.90	16.0	0.34	0.24	*LD*	0.84	0.41
8	5.5	*G*	1.00	*G*	1.00	*M*	0.56	*A*	1.00	*P*	0.90	12.6	0.44	0.12	*LD*	0.85	0.51
9	6.5	*L/H*	0.78	*G*	1.00	*N*	0.71	*A*	1.00	*P*	0.90	12.5	0.52	0.06	*LD*	1.00	0.52
10	6.5	*L/H*	0.78	*G*	1.00	*M*	0.56	*A*	1.00	*P*	0.90	9.8	0.66	0.03	*LD*	1.00	0.66
11	6.5	*G*	1.00	*G*	1.00	*N*	0.71	*A*	1.00	*P*	0.90	16.0	0.41	0.05	*LD*	1.00	0.41
12	6.5	*G*	1.00	*G*	1.00	*M*	0.56	*A*	1.00	*P*	0.90	12.6	0.52	0.02	*LD*	1.00	0.52
13	7.5	*L/H*	0.78	*G*	1.00	*N*	0.71	*A*	1.00	*P*	0.90	12.5	0.60	0.06	*LD*	1.00	0.60
14	7.5	*L/H*	0.78	*G*	1.00	*M*	0.56	*A*	1.00	*P*	0.90	9.8	0.76	0.03	*LD*	1.00	0.76
15	7.5	*G*	1.00	*G*	1.00	*N*	0.71	*A*	1.00	*P*	0.90	16.0	0.47	0.05	*LD*	1.00	0.47
16	7.5	*G*	1.00	*G*	1.00	*M*	0.56	*A*	1.00	*P*	0.90	12.6	0.60	0.02	*LD*	1.00	0.60
17	8.9	*L/H*	0.78	*G*	1.00	*N*	0.71	*A*	1.00	*P*	0.90	12.5	0.71	0.28	*LD*	0.84	0.85
18	8.9	*L/H*	0.78	*G*	1.00	*M*	0.56	*A*	1.00	*P*	0.90	9.8	0.91	0.14	*LD*	0.85	1.07
19	8.9	*G*	1.00	*G*	1.00	*N*	0.71	*A*	1.00	*P*	0.90	16.1	0.56	0.22	*LD*	0.85	0.66
20	8.9	*G*	1.00	*G*	1.00	*M*	0.56	*A*	1.00	*P*	0.90	12.5	0.71	0.11	*LD*	0.85	0.83

Note: *L/H* = low or high, *G* = good, *N* = near, *M* = medium, *F* = far, *A* = absent, *P* = poor, *LD* = long duration.

Depending on the data signed regarding geometries, individual FILI have been generated (four FILI in each weight category). They systematically present these combinations:

- V good and H near
- V good and H medium
- V bad and H near
- V bad and H medium

Frequencies have been estimated for each subtask and FILI, based on partial weight category frequency (previously determined) and the proportion of times each combination has been reported.

In the software this part is not visible because the subtasks simplification and FILI estimation are totally automatic.

In Exercise 5, to compute the final VLI, the traditional composite lifting index (CLI) equation cannot be applied correctly when there are as many as 20 subtasks (especially considering individual frequencies).

Generally speaking, unless there are fewer than 10 subtasks (and accordingly, 10 FILIs and 10 STLIs), the potential maximum 30 STLIs need to be grouped into less than 10 LI categories: In our proposal, it was decided to group them into six LI categories.

To generate these six LI categories, it is necessary to compute the FILI for each of the possible 30 subtasks previously defined. Considering the real distribution of all the resulting FILI values, the values corresponding to the sextiles (16.6th, 33.3th, 50th, 66.6th, and 83.3th percentiles) of the FILI distribution are selected; the sextiles' key values determine the limits for the six LI categories and the consequent grouping of the subtasks and relative FILI; the decision to use sextiles was designed to maximise the cumulated lifting frequencies in each category and to avoid empty categories. A simpler alternative that can be used for manually grouping FILI categories involves dividing the range of FILI values (from minimum to maximum) into six. However, this option does not optimise the cumulated frequencies for each category and can lead to empty FILI categories.

Once the FILI categories have been aggregated, a mean FILI value is chosen within each category, except for the last one (i.e., the one with the highest value), in which the highest FILI value is chosen.

Once the original (up to 30) FILI values are assigned to the appropriate LI category, it will also be possible to determine the cumulative lifting frequency for each of these six new LI categories; the frequency is generated by the sum of the frequencies associated with the individual FILI values grouped into each category. Using these cumulative frequencies, it is also possible to determine the corresponding frequency multipliers (FMs) based on traditional tables in the applications manual [Waters et al., 1994] considering the appropriate lifting duration scenario (short, medium, long).

At this point, the respective STLI values can be determined for each of these six new FILI categories (i.e., the category LI based on frequency). The STLI is defined as the LI value for each category, independent of the other categories.

The categories are renumbered in order of decreasing physical stress, beginning with the category with the greatest single task lifting index (STLI) down to the task category with the smallest STLI. The task categories are renumbered in this way so that the more difficult categories are considered first.

The VLI for the job is then computed according to the following formula (same formula as used for the CLI):

$$VLI = STLI_1 + \Sigma\Delta LI \tag{8.1}$$

where

$$\Sigma\Delta LI = \left(FILI_2 X \left(\frac{1}{FM_{1,2}} - \frac{1}{FM_1} \right) \right) + \left(FILI_3 X \left(\frac{1}{FM_{1,2,3}} - \frac{1}{FM_{1,2}} \right) \right) + \left(FILI_4 X \left(\frac{1}{FM_{1,2,3,4}} - \frac{1}{FM_{1,2,3}} \right) \right) + \left(FILI_n X \left(\frac{1}{FM_{1,2,3,4,\ldots,n}} - \frac{1}{FM_{1,2,3,\ldots,(n-1)}} \right) \right)$$

Note that the numbers in the subscripts refer to the new LI category numbers; the FM values are determined from the frequency table published in the applications manual [Waters et al., 1994].

Nonexperts in math should be reassured to know that the software automatically performs these complex calculations.

Coming back to Exercise 5, once the necessary data are entered, the software computes (sheet "5.LI") the final VLI (Table 8.9), according to international standard suggestions (broken down by gender and age) and to original NIOSH suggestions (non-gender specific and fixed load constant of 23 kg).

TABLE 8.9
Exercise 5: Variable Task: Final VLI Values According to International Standard Suggestions and to Original RNLE (see software sheet "5.LI")

European Standard (EN) 1005-2, ISO Standard 11228-1		
25 kg	Men (18–45 years old)	1.29
20 kg	Women (18–45 years old)	1.61
20 kg	Men (<18 or >45 years old)	1.61
15 kg	Women (<18 or >45 years old)	2.15
Revised NIOSH Lifting Equation (RNLE)		
23 kg	NIOSH criteria	1.40

8.3 ANOTHER EXAMPLE (EXERCISE 6) OF COMPUTATION OF A VARIABLE LIFTING TASK

8.3.1 General Aspects

In this section, another example (Exercise 6) of variable lifting task will be presented explaining more in detail all the analytical steps for computation, the relative criteria, and the applications of two mathematical models for partial computations within the same general approach that was previously presented. The first model is more detailed and complex and necessarily asks for a dedicated software; the second is somewhat easier when identifying subtasks and their partial frequencies and when aggregating the final six LI categories (these steps could be applied also by pen and paper but a software is indeed useful for performing all computational steps to obtain the final VLI).

The first three applicative steps (Parts A, B, and C) of the example (Exercise 6) are common to both computational models so they are described all together. Further steps are slightly different in the two mathematical models and will be described separately in Sections 8.3.2 and 8.3.3, respectively.

8.3.1.1 Part A: The Lifted Weights

In a warehouse a worker each day has to manually lift different loads, taking them from different trans-pallets and positioning them on shelves at different heights.

Figure 8.3A shows the different weights lifted and their number (as proposed in the specific software). In Figure 8.3B the different weights are divided (automatically by software) in five weight categories, using the following criteria (here exposed step by step) to obtain:

- The ranges of the new five categories
- The average weight load for each category that will be considered in the computation of the lifting index

The first step, to obtain the *five category ranges*, starts evaluating the indicated values:

- The minimum weight lifted: 4.5 kg (*mw*)
- The maximum weight lifted: 23.5 kg (*Mw*)
- The delta between the two values; having five categories is obtained with the following formula:

$$(Mw - mw)/5 = 3.8 \quad (8.2)$$

The range of the new five weight categories is, at the end, obtained, starting from the minimum weight value (4.5 kg) and adding to this value the delta (3.8 kg):

- From 4.5 (+3.8) to 8.29
- From 8.3 (+3.8) to 12.09
- From 12.1 (+3.8) to 15.89

	Representative Weight (kg)	Number of Objects to Lift per Shift by the Whole Group
From 4 to 4.99	4.5	400
From 8 to 8.99	8.5	300
From 13 to 13.99	13.5	100
From 16 to 16.99	16.5	80
From 20 to 20.99	20.5	40
From 23 to 23.99	23.5	5

Part A

Weight Categories		Number of Objects	Average weighted (Kg)	% Lifted Objects
From	To			
4.50	8.29	400.0	4.5	43.2%
8.30	12.09	300.0	8.5	32.4%
2.10	15.89	100.0	13.5	10.8%
15.90	19.69	80.0	16.5	8.6%
19.70	23.50	45.0	20.8	4.9%

- The minimum value is 4.5 kg (m).
- The maximum value is 23.5 (M).
- The delta between the two values having five categories is obtained with the following formula: $(M - m)/5 = 3.8$.

The five weight categories are, at the end, obtained, starting from the minimum weight value (4.5 kg) and adding to this value the calculated delta (3.8 kg):

- From 4.5 (+3.8) to 8.29
- From 8.3 (+3.8) to 12.09
- From 12.1 (+3.8) to 15.89
- From 15.9 (+3.8) to 19.69
- From 19.7 (+3.8) to 23.5

Part B

FIGURE 8.3 Exercise 6. Quantification of the different weights lifted (A) and the different weights divided into five new weight categories (B).

- From 15.9 (+3.8) to 19.69
- From 19.7 (+3.8) to 23.5

The second step allows the evaluation of the *average weight load* that will represent *each weight category*. It is obtained (Table 8.10) through the following procedure: For each new weight category the cumulative mass (CM) for each category is evaluated (see Table 8.10, column *c*) using the following formula:

$$CM = \Sigma(ai^*bi) \tag{8.3}$$

where *ai* is the representative weight for the 1 kg categories (column *a*) actually lifted, and *bi* is the number of objects lifted for the 1 kg categories (column *b*) actually lifted. Once the specific cumulative mass for each category is obtained, the specific average weight for the category is derived by dividing it by the number of objects present in its category (see column *e*).

To show how this step is accomplished, a second set of data is shown with different frequencies. In this second example (Table 8.11, Example A) the same number of pieces and the same weight categories are present, but with a different distribution of them inside each of the new five weight categories. Comparing the results of the two tables, it is possible to see the significant change of the average weight loads inside each of the new five weight categories.

Even in this example, again the weighted-average method is used to determine the average weight lifted, for each of the new five weight categories. For example, for the category from 4.5 to 8.29, it performs the following calculation:

$$[(4.5^*50) + (5.5^*350)]/400 = 5.37$$

8.3.1.2 Part B: The Job Organisation

Coming back to Exercise 6, we have a shift duration lasting 480 min: During this period practically all the time is spent in manual lifting.

Only few interruptions are present: two breaks of 15 min each and one lunch break of 30 min included in the shift duration (Table 8.12). The duration of manual lifting is considered the long duration.

8.3.1.3 Part C: Computing the Total (Overall) Frequency and the Partial Frequency (by Weight Categories) of Manual Lifting

Having a total shift duration of 480 min and considering the presence of 60 min of breaks without any other task, the *net duration of the manual lifting task* is 420 min.

Knowing that the total number of lifted weights is 925, the total lifting frequency (lifts per minute) is easy to calculate (Table 8.13): 925/420 = 2.2.

On the other hand, knowing how many objects are lifted within each of the five weight categories in the given period, it is also possible to compute the partial frequencies for each of them.

TABLE 8.10
Exercise 6: Evaluation of the Average Weight Load for Each Category

Weight (kg) (a)	Objects Lifted (b)	New Weight Categories (c)	Cumulative Mass (d)	Cumulative Mass for Each Category (e)	Number of Pieces (f)	Average Weight (g)
4.5	400	4.50–8.29	1,800	1,800	400.0	4.5
5.5						
6.5						
7.5						
8.5	300	8.30–12.09	2,550	2,550	300.0	8.5
9.5						
10.5						
11.5						
12.5		12.10–15.89		1,350	100.0	13.5
13.5	100		1,350			
14.5						
15.5						
16.5	80	15.90–19.69	1,320	1,320	80.0	16.5
17.5						
18.5						
19.5						
20.5	40	19.70–23.50	820	937.5	45.0	20.8
21.5						
22.5						
23.5	5		117.5			

[a] Weight of the lifted objects (kg)
[b] Number of objects really lifted per shift by the whole group
[c] Range of the new weight categories (kg)
[d] Cumulative Mass for Weight (kg); where $d_i = a_i * b_i$
[e] Cumulative Mass for New Weight Categories (kg); where $e_j = \Sigma a_i * b_i$
[f] Number of Pieces for the New Weight Categories
[g] Weighted Average weight loads for each of the New Weight Categories; where $g_j = e_j/f_j$

TABLE 8.11
Example A: Evaluation of the Average Weight Load for Each Category

Weight (kg) (a)	Objects Lifted (b)	New Weight Categories (c)	Cumulative Mass (d)	Cumulative Mass (e)	Number of Pieces (f)	Average Weight (g)
4.5	50.0	4.50–8.29	225	2,150	400.0	5.4 (g.1)
5.5	350.0		1,925			
6.5						
7.5						
8.5	50.0	8.30–12.09	425	2,800	300.0	9.3 (g.2)
9.5	250.0		2,375			
10.5						
11.5						
12.5	20.0	12.10–15.89	270	1,430	100.0	13.5 (g.3)
13.5	80.0		1,160			
14.5						
15.5						
16.5	20.0	15.90–19.69	330	1,380	80.0	17.3 (g.4)
17.5	60.0		1,050			
18.5						
19.5						
20.5	5.0	19.70–23.50	102.5	1,032.5	45.0	22.9 (g.5)
21.5	5.0		107.5			
22.5						
23.5	35.0		822.5			

[a] Weight of the lifted objects (kg)
[b] Number of objects really lifted per shift by the whole group
[c] Range of the new weight categories (kg)
[d] Cumulative Mass for Weight (kg); where $d_i = a_i * b_i$
[e] Cumulative Mass for New Weight Categories (kg); where $e_j = \Sigma a_i * b_i$
[f] Number of Pieces for the New Weight Categories
[g] Weighted Average weight loads for each of the New Weight Categories

1. [(4.5*50) + (5.5*350)]/400 = 5.4
2. [(8.5*50) + (9.5*250)]/300 = 9.3
3. [(12.5*20) + (13.5*80)]/100 = 13.5
4. [(16.5*20) + 17.5*60)]/80 = 17.3
5. [(20.5*5) + (21.5*5) + (23.5*35]/45 = 22.9

TABLE 8.12
Exercise 6: The Shift Duration and Task/Breaks Distribution

Task	Minutes	
Manual Lifting Task	30	
Other Task or Breaks	10	
Push-Pulling Task	15	
Other Task or Breaks		
Manual Lifting Task	30	
Other Task or Breaks	10	
Push-Pulling Task	15	
Other Task or Breaks		
Manual Lifting Task	30	
Other Task or Breaks	10	
Push-Pulling Task	15	
Other Task or Breaks		
Manual Lifting Task	30	
Other Tasks or Breaks	10	
Push-Pulling Task	15	
Other Task or Breaks (Lunch)	30	
Manual Lifting Task	30	
Other Tasks or Breaks	10	
Push-Pulling Task	15	
Other Task or Breaks		
Manual Lifting Task	30	
Other Task or Breaks	10	
Push-Pulling Task	15	
Other Task or Breaks	120	Shift end

TABLE 8.13
Exercise 6: Evaluation of the Total Lifting Frequency

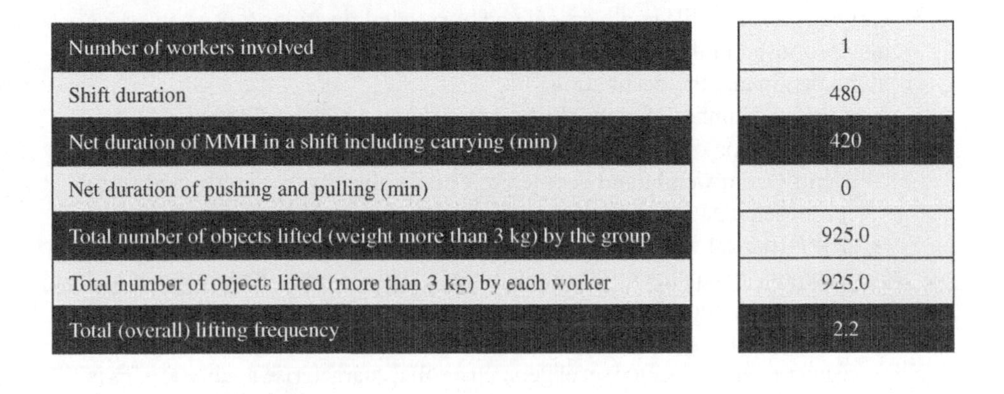

Long Duration	
Number of workers involved	1
Shift duration	480
Net duration of MMH in a shift including carrying (min)	420
Net duration of pushing and pulling (min)	0
Total number of objects lifted (weight more than 3 kg) by the group	925.0
Total number of objects lifted (more than 3 kg) by each worker	925.0
Total (overall) lifting frequency	2.2

TABLE 8.14
Exercise 6: Evaluation of Partial Weight Frequencies (PWFs) for Each Weight Category

Weight Category	Number of Objects	Lifting Period (min)	PWF (lpm)
1. From 4.5 to 8.29	400	420	0.95
2. From 8.3 to 12.09	300	420	0.71
3. From 12.1 to 15.89	100	420	0.24
4. From 15.9 to 19.69	80	420	0.19
5. From 19.7 to 23.5	45	420	0.11
All categories	925	420	2.20

In Exercise 6, in the 420 min period, we have the partial weight category frequencies reported (PWF) in Table 8.14.

8.3.2 Criteria and VLI Computation by a First Model (Exercise 6)

8.3.2.1 Part D: The Positions of the Weights at Origin and Destination (Geometries) per Weight Categories, the Subtasks, and Their Corresponding Partial Frequencies (PStFs)

It is now necessary to determine, for each weight category, the position of the weights (in relation to the body) at the origin and the destination to obtain the layout multipliers (vertical height, horizontal distance) and the presence of asymmetry.

This procedure will generate the subtasks to be analysed (up to a maximum of 30 potential subtasks), and a partial frequency (PStF) will be attributed to each of them.

Before starting, it is necessary to review some terminology and their meaning used in this context:

- *Vertical areas, horizontal areas*: The new simplified areas proposed in VLI computation that determine the corresponding vertical and horizontal multipliers.
- *Geometries*: The different vertical and horizontal positions (i.e., shelves), as also proposed in the software, where the lifted objects are kept or released at the origin and the destination.
- *Frequency*: Number of manual lifts per minute, which could be:
  - *TF* (total or overall frequency): Considers all lifted objects independently from weight and geometry: This is objectively determined from production data.
  - *PWF* (partial weight frequency): Considers all lifted objects in a weight category independently from geometry. This is objectively determined from production data.
  - *PStF* (partial subtask frequency): Considers all lifted objects in a weight category and a selected set of geometries that characterise a subtask. This is probabilistically determined from observation data on lifting geometries.

We start now with the description of the allocation of the objects present in the first weight category (4.5–8.3 kg) of Exercise 6, by steps.

1. First weight category: Consider the main general organisational data.
   - Number of pieces in this weight category = 400
   - Total net duration of manual lifting in the shift = 420 min
   - Frequency for this weight category (PWF) = 400/420 = 0.95 (rounded value)
2. First weight category: Report in the software (with *X*) the corresponding geometries of the lifted objects at origin and destination (Figure 8.4A).

3a. First weight category: evaluation of the total number of the geometries, occupied by objects, in the vertical and horizontal areas, considering the origin and destination (Figure 8.4B and Table 8.15) and subtasks definition.
   For this evaluation some criteria were adopted:
   a. All of the lifting tasks present in a variable task are considered to have significant control at the destination. Given that it is impossible to assess each individual lifting task, it was presumed that in most of them some kind of control (in the meaning reported in the NIOSH application manual [Waters et al., 1994]) could be present.
   b. Consequently we always consider horizontal and vertical positions at both the origin and the destination. The systematic computation of heights at the destination compensates (for variable tasks evaluation) for setting the vertical travel distance multiplier as a constant equal to 1 (practically a noninfluencing constant) (see Chapter 7).
   c. We consider the geometries at the origin and destination to obtain a more precise subtask frequency.

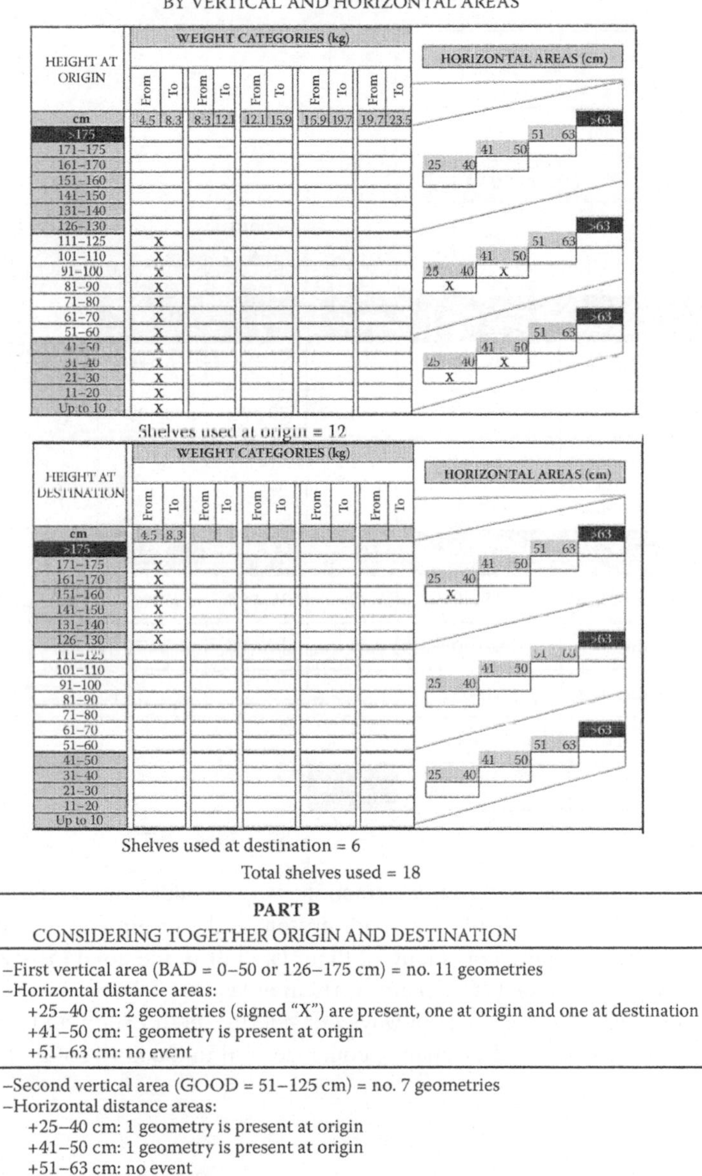

FIGURE 8.4 Exercise 6. Description of the geometries at origin and destination for the first weight category.

TABLE 8.15
Provisional Evaluation of Frequency of Each Subtask Only Considering the First Weight Category

Lifts in weight category (number of objects)	400				
Total net duration (min)	420				
Partial frequency	0.952				
	cm	Geometries			
		Number	%	Frequency	
Vertical area	0–50 or 126–175	11	61%	0.58	
Horizontal distance	25–40	2	40.74%	0.388	Subtask frequency 1
Horizontal distance	41–50	1	20.37%	0.194	Subtask frequency 2
Horizontal distance	51–63	0			
Horizontal distance	Total	3			
Vertical area	51–125	7	39%	0.37	
Horizontal distance	25–40	1	19.44%	0.185	Subtask frequency 3
Horizontal distance	41–50	1	19.44%	0.185	Subtask frequency 4
Horizontal distance	51–63				
Horizontal distance	Total	2			
Vertical area	Total	18			

d. For heights at the origin/destination, we consider how many vertical levels (by intervals of 10 cm) are present at the origin and the destination and compute how many of them (%) fall in the good (51–125 cm) and bad (0–50 or 126–175 cm) vertical areas.

e. For each of these areas (always considering both origin and destination), we compute how many geometries fall in the horizontal areas:
   - 25–40 cm
   - 41–50 cm
   - 51–63 cm

   In Exercise 6, for the first weight category, the geometries in vertical areas are (see Figure 8.4 and Table 8.15):
   - First vertical area (0–50 or 126–175 cm) = 11 geometries
   - Second vertical area (51–125 cm) = 7 geometries

   For each of the two vertical areas we can find objects in three different horizontal areas (Figure 8.4 and Table 8.15):

- First vertical area (0–50 or 126–175 cm) = 11 shelves
  +25–40 cm: 2 geometries (signed *X*) are present, 1 at origin and 1 at destination
  +41–50 cm: 1 geometry is present at origin
  +51–63 cm: no event
- Second vertical area (51–125 cm) = 7 shelves
  +25–40 cm: 1 geometry is present at origin
  +41–50 cm: 1 geometry is present at origin
  +51–63 cm: no event

Having two vertical areas and two horizontal areas, the presence of four subtasks results in the example (Table 8.15).

The descriptive numbers of geometries involved in horizontal and vertical areas are calculated. The goal of this calculation process is to obtain a proportional distribution of the objects in the subtasks present (4) in this weight category and consequently the PStF for each of them.

3b. First weight category: Estimation of the partial subtask lifting frequency (PStF). The procedure, using the example data in Table 8.15, is explained below.

Having gathered the data at the origin and destination, we observe, with respect to the vertical area:

- Eleven geometries are in bad areas (0–50 or 126–175 cm): This event means that 61% of the total 18 vertical geometries, occupied by weights, are in bad areas. These 11 vertical geometries are divided along horizontal distances:
  - Two are at a horizontal distance of 25–40 cm (2/3 by 61%) = 40.74% (subtask 1).
  - One is at a horizontal distance of 41–50 cm (1/3 by 61%) = 20.37% (subtask 2).
- Seven geometries are in good areas (51–125 cm) corresponding to 39% of the total 18 vertical geometries occupied by weights. These seven geometries are divided along horizontal distances:
  - One is at a horizontal distance of 25–40 cm (1/2 by 39%) = 19.44% (subtask 3).
  - One is at a horizontal distance of 41–50 cm (1/2 by 39%) = 19.44% (subtask 4).

Since the partial frequency of this weight category is 0.95, it becomes easy to calculate the representative frequency for each of the identified subtasks:

- (Subtask 1) = 40.74*0.95 = 0.388
- (Subtask 2) = 20.37*0.95 = 0.194
- (Subtask 3) = 19.44*0.95 = 0.185
- (Subtask 4) = 19.44*0.95 = 0.185

The subtask frequency can be calculated only when all the information regarding the adopted geometries (vertical heights, horizontal distances, eventual asymmetry) are reported for each weight category.

The procedure that has been detailed here for the first weight category has to be applied for all the other weight categories.

To this aim, in the previous example (Figure 8.2A and B) we showed that, depending on the observed pattern of horizontal distances, one may use two alternatives:

- The first (simpler; Figure 8.2B) is to consider horizontal distances the same for all the weight categories. This alternative should be used when horizontal distances are practically similar in all weight categories or it is impossible to distinguish in detail all the horizontal distances present in each of the five weight categories (see software ERGOepm_CLI&VLI_S).
- The second (more detailed; Figure 8.2A) is to consider horizontal distances separately for each weight category. This alternative should be used when horizontal distances are noticed to be different in the different weight categories or it is possible to distinguish in detail differences in horizontal distance (see software ERGOepm_VLI_AP).

In the following, both alternatives, applied to Exercise 6 (where only slight differences in horizontal distance among categories are reported), will be presented.

One should consider, however, that the difference between the two alternatives regards only the identification of horizontal distances, the consequent generation of subtasks, and the estimation of frequencies (PStFs).

Once subtasks have been identified and relative FILI values have been computed, the procedure for aggregating them in six LI categories and computing the VLI is exactly the same in the two computational alternatives. In this regard, at the end, the VLI results will be compared.

4. All the weight categories: Report in the software (with *X* or other symbols), in the corresponding geometries, the position of the lifted objects at the origin and destination (Figure 8.5).

   In Figure 8.5 all the geometries for all the weight categories are reported at the origin and destination.

   In Table 8.16 the evaluation process is reported that allows obtaining for each subtask the partial frequency (PStF), following the same procedure previously explained for the first weight category but extending it to the other weight categories.

   Considering that the horizontal areas are collectively described for all the weight categories together, the numbers of the geometries involved in the different horizontal areas appear the same for all the weight categories.

   In conclusion, it may appear difficult to understand how to obtain the partial frequency of each subtask. In fact the first approach, given the frequency for each of the five weight categories, was to equally divide it between the subtasks present in each weight category. But, having the opportunity to use the information regarding the position of the weights in the different geometries, it provides an idea of their *proportional distribution between the different subtasks identified within each weight category.*

The same Exercise 6 so far presented could be analysed (see ERGOepm_VLI_AP) based on a more detailed computational alternative that considers

OBJECTS DISTRIBUTION IN ALL THE DIFFERENT GEOMETRIES: THE VERTICAL FOR EACH WEIGHT CATEGORIES AND THE HORIZONTAL FOR ALL THE CATEGORIES TOGETHER

Geometries used at origin

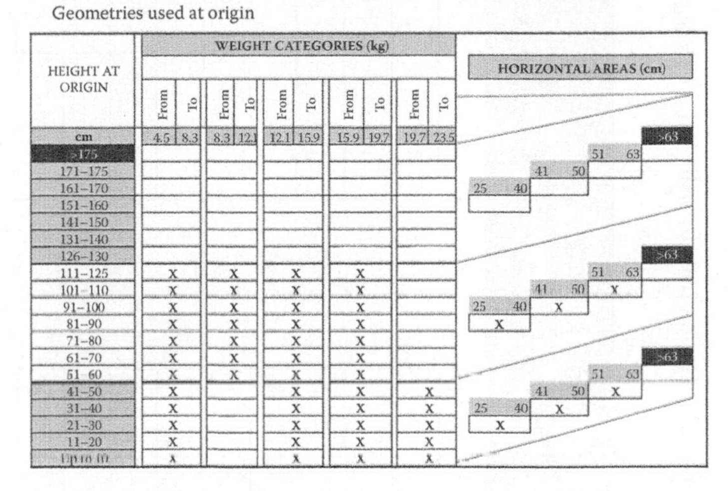

HEIGHT AT ORIGIN (cm)	WEIGHT CATEGORIES (kg) From 4.5 To 8.3	From 8.3 To 12.1	From 12.1 To 15.9	From 15.9 To 19.7	From 19.7 To 23.5
>175					
171–175					
161–170					
151–160					
141–150					
131–140					
126–130					
111–125	X	X	X	X	
101–110	X	X	X	X	
91–100	X	X	X	X	
81–90	X	X	X	X	
71–80	X	X	X	X	
61–70	X	X	X	X	
51–60	X	X	X	X	
41–50	X		X	X	X
31–40	X		X	X	X
21–30	X		X	X	X
11–20	X		X	X	X
Up to 10	X		X	X	X

HORIZONTAL AREAS (cm)

Height band	<25	25–40	41–50	51–63	>63
>175 to 126–130					
111–125 to 51–60	X	X	X		
41–50 to Up to 10	X	X	X		

Geometries used at destination

HEIGHT AT DESTINATION (cm)	WEIGHT CATEGORIES (kg) From 4.5 To 8.3	From To	From To	From To	From To
>175					
171–175	X				
161–170	X				
151–160	X				
141–150	X				
131–140	X				
126–130	X				
111–125		X	X	X	
101–110		X	X	X	
91–100		X	X	X	
81–90		X	X	X	
71–80		X	X	X	
61–70		X	X	X	
51–60		X	X	X	
41–50			X	X	X
31–40			X	X	X
21–30			X	X	X
11–20			X	X	X
Up to 10			X	X	X

HORIZONTAL AREAS (cm)

Height band	<25	25–40	41–50	51–63	>63
>175 to 126–130	X				
111–125 to 51–60	X				
41–50 to Up to 10	X				

ASYMMETRY

Asymmetry / Degrees	Weight Categories (Kg) From 4.5 To 8.3	From 8.3 To 12.1	From 12.1 To 15.9	From 15.9 To 19.7	From 19.7 To 23.5
More than 45° for More than 50% of the lifts					X
More than 135°					

FIGURE 8.5 Exercise 6. Vertical (analytically) and horizontal (synthetically) geometries reported for all weight categories at origin and destination.

TABLE 8.16

Exercise 6: Subtask Frequencies Obtained by the Simpler Method (Same Horizontal Geometries for All Weight Categories) (Total of 24 Subtasks)

	Average Weight Category (kg)	Number of Weights and Frequency for Weight Categories	Vertical Geometries (number and cm)	Horizontal Geometries (cm)	Final Number Horizontal Geometries	Subtasks	Asymmetry More Than 45° for More Than 50% of the Lifts	% of Objects for Each Subtask	Subtask Frequency
From 4.5 To 8.29	4.5	400	0–50 or 126–175	25–40	3	X	No	36.67%	0.349
			Number of geometries 11	41–50	1	X		12.22%	0.116
	Frequency	0.95		51–63	1	X		12.22%	0.116
			51–125	25-40	2	X		19.44%	0.185
			Number of geometries 7	41–50	1	X		9.72%	0.093
				51–63	1	X		9.72%	0.093
From 8.3 To 12.09	8.5	300	0–50 or 126–175	25–40			No		
			Number of geometries 0	41–50					
	Frequency	0.71		51–63					
			51–125	25–40	2	X		50.00%	0.357
			Number of geometries 14	41–50	1	X		25.00%	0.179
				51–63	1	X		25.00%	0.179

From 12.1 To 15.89	13.5	100	0–50 or 126–175	25–40	3	X	No	25.00%	0.060
			Number of geometries 10	41–50	1	X		8.33%	0.020
	Frequency	0.24		51–63	1	X		8.33%	0.020
			51–125	25–40	2	X		29.17%	0.069
			Number of geometries 14	41–50	1	X		14.58%	0.035
				51–63	1	X		14.58%	0.035
From 15.9 To 19.69	16.5	80	0–50 or 126–175	25–40	3	X	No	25.00%	0.048
			Number of geometries 10	41–50	1	X		8.33%	0.016
	Frequency	0.19		51–63	1	X		8.33%	0.016
			51–125	25–40	2	X		29.17%	0.056
			Number of geometries 14	41–50	1	X		14.58%	0.028
				51–63	1	X		14.58%	0.028
From 19.7 To 23.5	20.8	45	0–50 or 126–175	25–40	3	X	Yes	60.00%	0.064
			Number of geometries 10	41–50	1	X		20.00%	0.021
	Frequency	0.11		51–63	1	X		20.00%	0.021
			51–125	25–40					
			Number of geometries 0	41–50					
				51–63					
					Number total subtasks	24		Total frequency	2.2

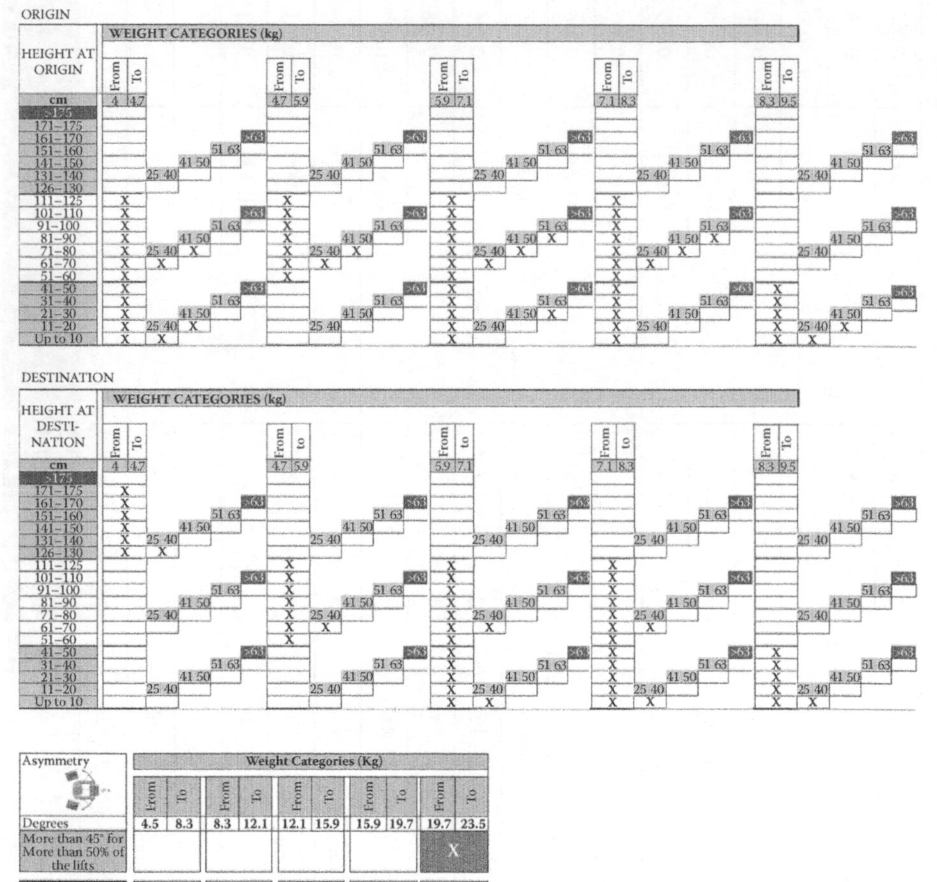

FIGURE 8.6 Exercise 6 BIS. Vertical and horizontal geometries analytically reported for all weight categories at origin and destination.

analytically horizontal geometries for each weight category (Exercise 6BIS, Figure 8.6).

Table 8.17 reports the analytical frequencies of the subtasks obtained by knowing the corresponding horizontal geometries, now relative to each weight category.

Comparing the two columns with data on the subtasks frequencies obtained by the two alternatives presented (Table 8.16, Exercise 6 and Table 8.17, Exercise 6BIS), one can see the small differences between the two results. The main difference regards the resulting subtasks (24 in Exercise 6 vs. 19 in Exercise 6BIS). This is due to a better identification of horizontal distances for each weight category by the second alternative.

Table 8.18 reports all the analytical data concerning the proposed Exercise 6BIS, regarding the 19 subtasks (derived from Table 8.17) and corresponding individual FILI and STLI.

Nineteen LI (for corresponding subtasks) are too many (especially considering the individual frequency of lifting) for correctly applying the traditional composite lifting index (CLI) approach [Waters et al., 1994]. So, it is suggested to proceed to another aggregation grouping the 19 individual LI into 6 LI categories.

In order to generate those six LI categories, the FILIs previously determined are to be considered.

The analytical procedure is illustrated in Table 8.19.

Among those 19 FILI values, the 16.66th, 33.33th, 50th, 66.66th, and 83.33th percentile values are determined. Those key percentiles take into account the variability of obtained results and determine the limits for aggregating the subtasks into six LI categories. Consequently the cumulative frequency of lifting for each of those six LI categories is also determined.

Once the FILI categories have been aggregated, a mean FILI value is chosen within each category, except for the last one (i.e., the one with the highest values) in which the highest FILI value is chosen.

Please note how, using the sextile technique, the 19 FILI values are quite equally distributed into the six generated LI categories.

One may now compute the final VLI by using the traditional CLI formula applied to the six LI categories. Table 8.20 reports all the elements useful to calculate the final VLI for the new six LI.

Using the data of Table 8.20, organised in six LI categories, it will be possible to compute the VLI by means of the traditional CLI formula:

$$VLI = STLI_1 + \Sigma\Delta LI \tag{8.4}$$

Based on the data presented for Exercise 6BIS, the computational data for applying the formula are reported in Table 8.21.

Using data reported in Table 8.21, the VLI for Exercise 6BIS can be calculated as follows:

$$VLI = STLI_1 + \Sigma\Delta LI$$

$$STLI_1 = 4.303$$

$$\Delta FILI_2 = 1.752*[(1/0.841) - (1/0.85)] = 1.752*(0.013) = 0.023$$

$$\Delta FILI_3 = 1.239*[(1/0.825) - (1/0.841)] = 1.239*(0.023) = 0.029$$

$$\Delta FILI_4 = 0.983*[(1/0.804) - (1/0.825)] = 0.983*(0.031) = 0.031$$

$$\Delta FILI_5 = 0.555*[(1/0.704) - (1/0.804)] = 0.555*(0.177) = 0.098$$

$$\Delta FILI_6 = 0.333*[(1/0.627) - (1/0.704)] = 0.363*(0.174) = 0.058$$

Hence:

$$VLI = STLI_1 + \Delta FILI_2 + \Delta FILI_3 + \Delta FILI_4 + \Delta FILI_5 + \Delta FILI_6$$

$$VLI = 4.303 + 0.023 + 0.029 + 0.031 + 0.098 + 0.058 = 4.54$$

TABLE 8.17
Exercise 6BIS: Subtask Frequencies Obtained with Analytical Method That Considers Horizontal Geometries for Each Weight Category (19 Subtasks)

	Average Weight Category (kg)	Number of Weights and Frequency for Weight Categories	Vertical Geometries (number and cm)	Horizontal Geometries (cm)	Final Number Horizontal Geometries	Subtasks	Asymmetry More Than 45° for More Than 50% of the Lifts	% of Objects for Each Subtask	Subtask Frequency
From 4.5 To 8.29	4.5	400	0–50 or 126–175	25–40	2	X	No	40.741%	0.388
			Number of geometries 11	41–50	1	X		20.370%	0.194
	Frequency	0.95		51–63					
			51–125	25–40	1	X		19.444%	0.185
			Number of geometries 7	41–50	1	X		19.444%	0.185
				51–63					
From 8.3 To 12.09	8.5	300	0–50 or 126–175	25–40					
			Number of geometries 0	41–50					
	Frequency	0.71		51–63					
			51–125	25–40	2	X		66.667%	0.476
			Number of geometries 14	41–50	1	X		33.333%	0.238
				51–63					

From 12.1 To 15.89	13.5	100	0–50 or 126–175	25–40	1	X	No	20.833%	0.050
			Number of geometries 10	41–50					
	Frequency	0.24		51–63	1	X		20.333%	0.050
			51–125	25–40	2	X		29.167%	0.069
			Number of geometries 14	41–50	1	X		14.583%	0.035
				51–63	1	X		14.583%	0.035
From 15.9 To 19.69	16.5	80	0–50 or 126–175	25–40	1	X	No	20.833%	0.040
			Number of geometries 10	41–50					
	Frequency	0.19		51–63	1	X		20.833%	0.040
			51–125	25–40	2	X		29.167%	0.056
			Number of geometries 14	41–50	1	X		14.583%	0.028
				51–63	1	X		14.583%	0.028
From 19.7 To 23.5	20.8	45	0–50 or 126–175	25–40	1	X	Yes	33.333%	0.036
			Number of geometries 10	41–50	1	X		33.333%	0.036
	Frequency	0.11		51–63	1	X		33.333%	0.036
			51–125	25–40					
			Number of geometries 0	41–50					
				51–63					
					Number total subtasks	19		Total frequency	2.2

TABLE 8.18

Exercise 6BIS: Analytical Data Regarding the 19 Identified Subtasks, the Relative Multipliers and Corresponding Individual FILI and LI per Each of the 19 Subtasks (FILI and LI computed for a reference mass of 25 kg)

Subtasks	Weight (kg)	Vertical Areas		Horizontal Areas		Asymmetry		Grasp		FILI	Frequency	Duration		LI
1	4.5	0–50 or 126–175 cm	0.78	25–40 cm	0.71	No	1.00	*p*	0.90	0.36	0.39	Long	0.82	0.55
2	4.5	0–50 or 126–175 cm	0.78	41–50 cm	0.56	No	1.00	*p*	0.90	0.46	0.19	Long	0.85	0.67
3	4.5	51–125 cm	1.00	25–40 cm	0.71	No	1.00	*p*	0.90	0.28	0.19	Long	0.85	0.41
4	4.5	51–125 cm	1.00	41–50 cm	0.56	No	1.00	*p*	0.90	0.36	0.19	Long	0.85	0.53
5	8.5	51–125 cm	1.00	25–40 cm	0.71	No	1.00	*p*	0.90	0.53	0.48	Long	0.81	0.82
6	8.5	51–125 cm	1.00	41–50 cm	0.56	No	1.00	*p*	0.90	0.67	0.24	Long	0.84	1.00
7	13.5	51–125 cm	0.78	25–40 cm	0.71	No	1.00	*p*	0.90	1.08	0.05	Long	1.00	1.35
8	13.5	51–125 cm	0.78	51–63 cm	0.40	no	1.00	*p*	0.90	1.92	0.05	Long	1.00	2.40
9	13.5	0–50 or 126–175 cm	1.00	25–40 cm	0.71	No	1.00	*p*	0.90	0.85	0.07	Long	1.00	1.06
10	13.5	0–50 or 126–175 cm	1.00	41–50 cm	0.56	No	1.00	*p*	0.90	1.07	0.03	Long	1.00	1.34
11	13.5	0–50 or 126–175 cm	1.00	51–63 cm	0.40	No	1.00	*p*	0.90	1.50	0.03	Long	1.00	1.88
12	16.5	51–125 cm	0.78	25–40 cm	0.71	No	1.00	*p*	0.90	1.32	0.04	Long	1.00	1.66
13	16.5	51–125 cm	0.78	51–63 cm	0.40	No	1.00	*p*	0.90	2.35	0.04	Long	1.00	2.94
14	16.5	0–50 or 126–175 cm	1.00	25–40 cm	0.71	No	1.00	*p*	0.90	1.03	0.06	Long	1.00	1.29
15	16.5	0–50 or 126–175 cm	1.00	41–50 cm	0.56	No	1.00	*p*	0.90	1.31	0.03	Long	1.00	1.64
16	16.5	0–50 or 126–175 cm	1.00	51–63 cm	0.40	No	1.00	*p*	0.90	1.83	0.03	Long	1.00	2.29
17	20.8	51–125 cm	0.78	25–40 cm	0.71	No	1.00	*p*	0.90	2.06	0.04	Long	1.00	2.58
18	20.8	51–125 cm	0.78	41–50 cm	0.56	No	1.00	*p*	0.90	2.61	0.04	Long	1.00	3.27
19	20.8	51–125 cm	0.78	51–63 cm	0.40	No	1.00	*p*	0.90	3.66	0.04	Long	1.00	4.57

By using the more detailed (with respect to horizontal distances) computational alternative, the final VLI for Exercise 6BIS is equal to 4.54 (considering a reference mass of 25 kg) or to 4.94 if one considers a reference mass of 23 kg.

We may consider now, following the same logic, use of the simpler alternative for assessing VLI starting from data reported in Table 8.16 (Exercise 6).

Table 8.22 reports the results of the calculation of the final VLI obtained by using the two different ways: the simpler (alternative 1, Exercise 6) and the detailed (alternative 2, Exercise 6BIS).

In Exercise 6 the resulting subtasks were 24; in Exercise 6BIS the resulting subtask were less (19) due to a better identification of horizontal distances for each weight category. However, the final results (VLI) are practically the same (4.54 vs. 4.56) when using the two alternatives.

This similarity of results, among others, is also due to the consideration that:

- Both of the alternatives perform the same procedure reducing the subtasks' number (24 and 19, respectively) to six LI categories.
- The final formula calculation of the VLI, using the same criteria proposed for the CLI, favours the worst LI category (STLI Max), weighting it for the LI of other LI categories. In the present example, STLI Max is practically the same for the two alternatives, thus leading to similar results.

In practice, two alternatives are proposed to estimate the VLI by this model 1: They have the only difference to use a different precision for describing the horizontal distances in respect to the weight categories. The final results are basically the same.

We leave open to users the option between the two models with the suggestion to choose the most appropriate option, given the different descriptive needs regarding the horizontal geometries of the objects handled.

In particular these suggestions are given:

- The first (simpler) alternative should be used when horizontal distances are practically similar in all weight categories or it is impossible to distinguish in detail all the horizontal distances present in each of the weight categories (see software ERGOepm_CLI&VLI_S).
- The second (more detailed) alternative should be used when horizontal distances are noticed to be different in the different weight categories or it is possible to distinguish in detail differences concerning each of the weight categories (see software ERGOepm_VLI_AP).

Exercise 5 was a clear example of a preferential use of the first, simpler, alternative.

Exercises 6 and 6BIS represent an intermediate condition where both alternatives could be used. However, VLI results are expected to be similar.

Exercise 7 (see later) will present an example for a preferential use of the second, more detailed, alternative.

TABLE 8.19
Exercise 6BIS: Procedure for Aggregating the Subtasks into Six STLI Categories Using Key Percentiles (Sextiles)

	6 FILI Categories Using Corresponding %ILE					
Category	**1**	**2**	**3**	**4**	**5**	**6**
Representative FILI value	2.06	1.50	1.08	0.84	0.46	0.28
Percentile	>83%	>67%	>50%	>33%	>17%	>0%
Subtasks						
1						X
2					X	
3						X
4						X
5					X	
6					X	
7			X			
8		X				
9				X		
10				X		
11		X				
12			1.324			
13	X					
14				X		
15			X			
16		X				
17	X					
18	X					
19	X					
Frequencies	0.16	0.11	0.12	0.16	0.91	0.77

TABLE 8.20
Exercise 6BIS: Relevant Values for Each of the Six STLI Categories Useful for Final VLI Computation (considering a reference mass of 25 kg)

	1	2	3	4	5	6
STLI max.	3.66	1.92	1.32	1.07	0.67	0.36
STLI average	2.67	1.75	1.24	0.98	0.55	0.33
Frequency	0.16	0.11	0.12	0.16	0.91	0.77
Duration	Long	Long	Long	Long	Long	Long
F multiplier	0.85	0.85	0.85	0.85	0.76	0.78
LI max.	4.304	2.262	1.558	1.26	0.882	0.463

TABLE 8.21
Exercise 6BIS: Relevant Data for Computing Final VLI Derived from Table 8.20

Connotation of Cumulative Frequencies by STLI Order	Cumulative Frequencies of Categories (lpm)	Corresponding FM (long duration) (*A*)	Partial Value $[(1/FM_i) - (1/FM_{i-1})]$ (*B*)	FILI (*C*)	$STLI_1$ $(C \times A)$ $\Delta FILI_i$ $(C \times B)$
FM_1	0.16	0.850		3.658	4.303
$FM_{1,2}$	0.27	0.841	0.013	1.752	0.023
$FM_{1,2,3}$	0.39	0.825	0.023	1.239	0.029
$FM_{1,2,3,4}$	0.55	0.804	0.031	0.983	0.031
$FM_{1,2,3,4,5}$	1.46	0.704	0.177	0.555	0.098
$FM_{1,2,3,4,5,6}$	2.20	0.627	0.174	0.333	0.058

8.3.3 Criteria and VLI Computation by a Second Model (Exercise 6TER)

As introduced in Section 8.3.1, a second model for computing VLI, in the framework of the same general approach, could be used.

This second model is somewhat easier when identifying subtasks and their partial frequencies and when aggregating the final six LI categories (these steps could also be applied by pen and paper but an application software is indeed useful for performing all computational steps to obtain the final VLI).

It differs from the previous one substantially in the procedures for identifying relevant geometries and subtasks, correspondent partial subtask frequencies, and consequent individual FILIs (up to 30). Once FILIs and their partial frequencies have been estimated, then the procedure to calculate the final VLI is similar to that in the previous model.

TABLE 8.22
Exercise 6 (24 Subtasks) and Exercise 6BIS (19 Subtasks): Comparison of Individual LI and the Final VLI (based on a reference mass of 25 kg)

Subtasks	Frequency Exercise 6	Frequency Exercise 6BIS	Duration Multiplier Exercise 6	Duration Multiplier Exercise 6BIS	LI Exercise 6	LI Exercise 6BIS
1	0.35	0.39	0.83	0.82	0.44	0.55
2	0.12	0.19	0.85	0.85	0.54	0.67
3	0.12	0.19	0.85	0.85	0.75	0.41
4	0.19	0.19	0.85	0.85	0.33	0.53
5	0.09	0.48	1.00	0.81	0.36	0.82
6	0.09	0.24	1.00	0.84	0.50	1.00
7	0.36	0.05	0.83	1.00	0.64	1.35
8	0.18	0.05	0.85	1.00	0.79	2.40
9	0.18	0.07	0.85	1.00	1.11	1.06
10	0.06	0.03	1.00	1.00	1.08	1.34
11	0.02	0.03	1.00	1.00	1.37	1.88
12	0.02	0.04	1.00	1.00	1.92	1.66
13	0.07	0.04	1.00	1.00	0.85	2.94
14	0.03	0.06	1.00	1.00	1.07	1.29
15	0.03	0.03	1.00	1.00	1.50	1.64
16	0.05	0.03	1.00	1.00	1.32	2.29
17	0.02	0.04	1.00	1.00	1.68	2.58
18	0.02	0.04	1.00	1.00	2.35	3.27
19	0.06	0.04	1.00	1.00	1.03	4.57
20	0.03		1.00		1.31	
21	0.03		1.00		1.83	
22	0.06		1.00		2.06	
23	0.02		1.00		2.61	
24	0.02		1.00		3.66	
					4.56	**4.54**

In general this second model uses the aggregation of lifted objects into up to five weight categories and the simplifications that have been presented regarding geometries (Chapter 7).

Based on these simplifications, a FILI can be calculated for each possible combination of lifting variables (H = horizontal, V = vertical, and A = asymmetry, at the origin and destination of the lift).

Before the FILI can be calculated for each possible combination of H and V, it should be determined if significant control is required [Waters et al., 1994].

If significant control is required, then the worst-case combination of factors at the origin and destination of the lifts will be used for that possible combination of factors. Also, in order to simplify the assessment, the decision regarding asymmetry is made for each weight category. Once the asymmetry has been chosen, the load constant is reduced for every possible combination of H and V according to the decision on asymmetry. After the asymmetry factor is applied, the FILI for each possible combination of H and V values at the origin and destination of the lift can be calculated and the overall distribution of possible combinations can be determined.

As can be seen below, for each possible combination of V and H, the following submultipliers can be determined, in order of magnitude:

GN (V = good, H = near) = 1.0*0.71 = 0.71
GM (V = good, H = mid) = 1.0*0.56 = 0.56
BN (V = bad, H = near) = 0.78*0.71 = 0.5538 (rounded to 0.55)
BM (V = bad, H = mid) = 0.78*0.56 = 0.4368 (rounded to 0.44)
GF (V = good, H = far) = 1.0*0.40 = 0.40
BF (V = bad, H = far) = 0.78*0.40 = 0.312 (rounded to 0.31)

For any combination of factors at the origin and destination of the lift (formally a subtask or subgroup), only one of the six categories would apply to any specific lift. For example, for the combination of GN at the origin and BM at the destination (GN-BM), the lift would be categorised into the BM category because the BM submultiplier (0.44) is lower than the GN category (0.71). Similarly, for the combination GF-BM, the lift would be categorised into the GF category because the GF submultiplier (0.40) is lower than the BM value (0.44).

If significant control is not required for the lifts (i.e., significant control must not be required for more than 80% of the lifts to apply), then you only have to consider the H and V combinations at the origin to determine the distributions.

It should be noted, however, that based on applicative experiences, in most cases where variable tasks are performed, significant control for most of the lifts is often required. This leads to consideration of both origin and destination. Moreover, since in the VLI approach the distance multiplier (DM) is considered a constant equal to 1, it is suggested to consider systematically the geometries at both the origin and the destination for a better risk estimation.

Also, it should be noted that this procedure assumes an equal distribution of lifts from the various possible H and V combinations at the origin and destination of the lifts, which will not always be the case. If the variability is too great, then a sampling approach or the method previously presented (Section 8.3.2) may be more applicable.

This model will be now applied to Exercise 6TER.

The first three applicative steps (Parts A, B, and C) of the example (Exercise 6) are common to the previous model and have been described in Section 8.3.1. Now the following steps applying the model will be reported.

8.3.4 Exercise 6TER (Second Model)

8.3.4.1 Part D: The Positions of the Weights at the Origin and Destination (Geometries) per Weight Categories, the Subtasks, and Their Corresponding Partial Frequencies (PStFs)

Consider Figure 8.6 as a starting point.

In Figure 8.6 vertical and horizontal geometries are analytically reported for all weight categories at origin and destination. Moreover, consider that the weight category 19.7–23.5 kg presents asymmetry for most of lifts; *AM* = 0.81 will be applied for all the lifts in this weight category.

Significant control is present for all the lifting actions. A coupling multiplier was considered always poor (*CM*=0.9).

Computations will be based on a load constant of 23 kg. Partial data will be shown as rounded to two (for multipliers) or three (for frequencies of lifts and others) decimals.

Based on data of Figure 8.6, the following possibilities exist for lifts, by weight category.

8.3.4.1.1 Weight Category 4.5–8.3 kg

At the origin the VH combination is either GN, GM, BN, or BM; at the destination, the VH combination can only be BN. Therefore the possible combinations of V and H at origin and destination are GN-BN, GM-BN, BN-BN, and BM-BN. Consider now *HM* = horizontal multiplier and *VM* = vertical multiplier, 23 kg as the load constant and 4.5 kg as representative load for the category.

Examining these combinations and selecting the worst case (*HM***VM* value) for each combination of H and V results in ¾ BN and ¼ BM. The FILI value for the BN subtask is 0.393 = (4.5/23*0.55*0.9), and for the BM subtask the FILI value is 0.498 = (4.5/23*0.44*0.9). The partial frequencies for each of these subtasks can be calculated by multiplying the percentage of lifts for each subtask times the overall frequency for this weight category (i.e., *BNpF* = 0.75*0.952 = 0.714 lifts/min (lpm); *BMpF* = 0.25*0.952 = 0.238 lpm).

8.3.4.1.2 Weight Category 8.3–12.1 kg

At the origin, *V* = *G* or *B* and *H* = *N* or *M*; at the destination, *V* = *G* and *H* = *N*. Therefore the possible combinations of V and H at origin and destination are GN-GN and GM-GN. Examining these combinations and selecting the worst case (*HM***VM* value) for each combination results in 1/2 GN and 1/2 GM. For this weight category (8.3–12.1), the FILI value for the GN subtask is 8.5/23*0.71*0.9 = 0.578 and the FILI value for the GM subtask is 8.5/23*0.56*0.9 = 0.733. The partial frequencies for each of these subtasks can be calculated by multiplying the percentage of lifts for each subtask times the overall frequency for this weight category (i.e., *GNpF* = 0.5*0.714 = 0.357 lpm and *GMpF* = 0.5*0.714 = 0.357 lpm).

8.3.4.1.3 Weight category 12.1–15.9 kg

At the origin, the VH combinations can be GN, GM, GF, or BF; at the destination, the VH combinations can be GN or BN. Therefore the possible combinations of V and H at the origin and destination are GN-GN, GN-BN, GM-GN, GM-BM, GF-GN, GF-BN, BF-GN, and BF-BN. Examining these combinations and selecting the worst case (*HM***VM* value) for each combination results in 1/8 GN, 1/8 GM, 2/8 BN, 2/8 GF, and 2/8 BF. For this weight category (12.1–15.9), the FILI values for the GN, GM, BN, GF, and BF subtask are 0.919 (13.5/23*0.71*0.9), 1.165 (13.5/23*0.56*0.9), 1.178 (13.5/23*0.55*0.9), 1.63 (13.5/23*0.4*0.9), and 2.09 (13.5/23*0.31*0.9), respectively. The partial frequencies for each of these subtasks can be calculated by multiplying the percentage of lifts for each subtask times the overall frequency for this weight category (i.e., *GNpF* = 0.125*0.238 = 0.03 lpm, *GMpF* = 0.125*0.238 = 0.03 lpm,

BNpF = 0.25*0.238 = 0.06 lpm, *GFpF* = 0.25*0.238 = 0.06, and *BFpF* = 0.25*0.238 = 0.06 lpm).

8.3.4.1.4 Weight Category 15.9–19.7 kg

The possible VH combinations for this weight category are the same as for the 12.1–15.9 category. At the origin, the VH combinations can be GN, GM, GF, or BF; at the destination, the VH combinations can be GN or BN. Therefore the possible combinations of V and H at origin and destination are GN-GN, GN-BN, GM-GN, GM-BM, GF-GN, GF-BN, BF-GN, and BF-BN. Examining these combinations and selecting the worst case (*HM**VM* value) for each combination results in 1/8 GN, 1/8 GM, 2/8 BN, 2/8 GF, and 2/8 BF. For this weight category (15.9–19.7 kg), the FILI values for the GN, GM, BN, GF, and BF subtasks are 1.123 (16.5/23*0.71*0.9), 1.423 (16.5/23*0.56*0.9), 1.439 (16.5/23*0.55*0.9), 1.993 (16.5/23*0.4*0.9), and 2.555 (16.5/23*0.31*0.9), respectively. The partial frequencies for each of these subtasks can be calculated by multiplying the percentage of lifts for each subtask times the overall frequency for this weight category (i.e., *GNpF* = 0.125*0.19 = 0.024 lpm, *GMpF* = 0.125*0.19 = 0.024 lpm, *BNpF* = 0.25*0.19 = 0.048 lpm, *GFpF* = 0.25*0.19 = 0.048 lpm, and *BFpF* = 0.25*0.19 = 0.048 lpm).

8.3.4.1.5 Weight Category 19.7–23.5 kg

At the origin, the VH combinations can be BM or BF; at the destination, the VH combination can only by BN. Therefore, the possible combinations of V and H at the origin and destination can be BM-BN and BF-BN. Examining these combinations and selecting the worst case (*HM**VM* value) for each combination results in ½ BM and ½ BF. For this weight category (19.7–23.5), the FILI values should consider also the presence of asymmetry (*AM* = 0.81) for all the lifts; hence the FILI values for BM and BF subtasks are 2.84 (20.8/23*0.44*0.9*0.81) and 3.976 (20.8/23*0.31*0.9*0.81), respectively. The partial frequencies for each of these subtasks can be calculated by multiplying the percent of lifts for each subtask by the overall frequency for this weight category (i.e., *BMpF* = 0.5*0.107 = 0.054 lpm and *BFpF* = 0.5*0.107 = 0.054).

The results are summarised in Table 8.23 for all weight categories and combinations of H and V at the origin and destination. These values can then be categorised into six LI categories and the VLI can be calculated.

The six LI categories may be derived by the sextile approach or by dividing the difference (FILI Max – FILI Min) in six equal parts.

In this case the six equal size categories was used.

Once the FILI categories have been aggregated, a mean FILI value is chosen within each category, except for the last one (i.e., the one with the highest values) in which the highest FILI value is chosen.

Table 8.24 reports all relevant data of the aggregation in the final six LI categories.

Based on the previous data (Table 8.24), the computational data for applying the final formula for VLI are reported in Table 8.25.

It should be noted that the value for the third category ($\Delta FILI_3$) equals zero since it resulted in a very low cumulated frequency in that category: This is a partial disadvantage deriving from using the (Max – Min)/6 approach instead of the sextile approach (see model 1) that usually maximises the cumulated frequencies within the categories.

TABLE 8.23
Exercise 6TER: Individual FILI (n = 16) and Corresponding Frequencies as Computed with Second Model (values are computed considering a load constant of 23 kg)

Resulting FILI	Estimated Frequency (lpm)	New LI Category (Table 8.24)
0.393	0.714	1
0.498	0.238	1
0.578	0.357	1
0.733	0.357	1
0.919	0.030	1
1.165	0.030	2
1.178	0.060	2
1.630	0.060	3
2.090	0.060	3
1.123	0.024	2
1.423	0.024	2
1.439	0.048	2
1.993	0.048	3
2.555	0.048	4
2.840	0.054	5
3.976	0.054	6

TABLE 8.24
Exercise 6TER: Results (from Table 8.23) for the Aggregation in Six Final LI Categories by the Second Model and by the (Max – Min)/6 Criterion for Individuating Category Limits (values are computed considering a load constant of 23 kg)

	LI Category ((3.976 – 0.393)/6 = 0.597)					
Category Data	**0.393–0.98**	**0.99–1.58**	**1.59–2.17**	**2.18–2.77**	**2.78–3.37**	**3.38–3.976**
Rep FILI within cat	0.624	1.266	1.904	2.555	2.840	3.976
Total F within cat (rounded)	1.7	0.18	0.17	0.05	0.05	0.05
$STLI_{max}$ within cat	1.351	1.693	2.459	2.555	2.840	3.976
Renumbered by $STLI_{max}$	6	5	4	3	2	1
FM	0.68	0.85	0.85	1	1	1

TABLE 8.25
Exercise 6TER: Relevant Data for Computing Final VLI Derived from Table 8.24 for the Case Study (values are computed considering a load constant of 23 kg)

Connotation of Cumulative Frequencies by STLI Order	Cumulative Frequencies of Categories (lpm)	Corresponding FM (long duration) (*A*)	Partial Value $[(1/FM_j) - (1/FM_{j-1})]$ (*B*)	FILI (*C*)	$STLI_1$ $(C \times A)$ $\Delta FILI_j$ $(C \times B)$
FM_1	0.05	1		3.976	3.976
$FM_{1,2}$	0.1	0.85	0.176	2.840	0.501
$FM_{1,2,3}$	0.15	0.85	0	2.555	0
$FM_{1,2,3,4}$	0.32	0.834	0.023	1.904	0.043
$FM_{1,2,3,4,5}$	0.50	0.81	0.036	1.266	0.045
$FM_{1,2,3,4,5,6}$	2.20	0.63	0.353	0.624	0.220

In conclusion, by using data reported in Table 8.25, the VLI for this variable lifting task (Exercise 6TER) can be calculated as follows:

$$VLI = STLI_1 + \Sigma\Delta LI \tag{8.4}$$

$$STLI_1 = 3.976$$

$$\Delta FILI_2 = 2.840*[(1/.85) - (1/1)] = 2.840*(0.176) = 0.501$$

$$\Delta FILI_3 = 2.555*[(1/.85) - (1/.85)] = 2.555*(0.0) = 0.0$$

$$\Delta FILI_4 = 1.904*[(1/.834) - (1/.85)] = 1.904*(0.023) = 0.043$$

$$\Delta FILI_5 = 1.266*[(1/.81) - (1/.834)] = 1.266*(0.036) = 0.045$$

$$\Delta FILI_6 = 0.624*[(1/.63) - (1/.81)] = 0.624*(0.356) = 0.220$$

Hence:

$$VLI = STLI_1 + \Delta FILI_2 + \Delta FILI_3 + \Delta FILI_4 + \Delta FILI_5 + \Delta FILI_6$$

$$VLI = 3.976 + 0.501 + 0.0 + 0.043 + 0.045 + 0.220 = 4.785$$

The final VLI, resulting by applying the second model to Exercise 6, is 4.79 (based on a load constant of 23 kg); if a load constant of 25 kg is applied, the final VLI, by this model, would be equal to 4.41.

This result is very similar to the ones (4.54) derived from the application of the first model to the same exercise (based on a load constant of 25 kg).

This similarity of results, among others, is also due to the consideration that:

- The two models have and use the same general framework and approach.
- Differences are confined to minor details regarding the way of estimating subtasks, respective individual frequencies, and FILI, and to less extent, to the procedures for the aggregation of final LI categories.
- In the present exercise, the relevant value of $STLI_1$ (in both computational models), derived from handling heavy loads in bad geometries (fifth weight category), greatly influences, in a similar way, the final VLI.

8.4 AN EXAMPLE OF VARIABLE TASK ANALYTICAL ANALYSIS IN AN ASSEMBLY LINE (EXERCISE 7)

8.4.1 General Aspects (Exercise 7)

This third example (Exercise 7) regards a variable manual lifting task performed at an assembly line. It has been selected since, in this case, different horizontal distances per weight categories are clearly identified at both the origin and the destination, thus leading to using the more analytical models that have been previously presented.

The example is based on original RNLE and its reference mass as suggested by NIOSH (23 kg).

At an assembly line one worker loads and unloads different boxes during an 8 h shift. The task is organised in cycles; during each cycle the worker handles all the kind of boxes in different body postures due to different heights (of the hands) at the origin and destination and different horizontal distances.

The shift lasts 480 min (from 8 a.m. to 5 p.m.). There is a break of 10 min at 10 a.m., lunchtime is at 12:10 a.m. (it lasts 60 min, out of official working time). In the afternoon the activity is the same as in the morning with a 10 min break at 3:10 p.m. and the last 40 min devoted to light work (no manual handling). Hence the total manual handling duration during the shift is 420 min.

Table 8.26 shows the sequence of lifting task, breaks, and light work during the shift.

The boxes have three different weights (6, 8, and 13 kg); the respective number of boxes lifted during the shift is shown in Table 8.27.

Since 1,852 boxes are lifted during a 420 min period, the overall lifting frequency is 4.41 lifts per minute (lpm). The partial lifting frequencies for each type (weight) of box are 1.18 lpm for 6 kg boxes, 2.94 lpm for 8 kg boxes, and 0.29 lpm for 13 kg boxes.

The duration scenario is a long duration type (continuative period of manual handling of 120 min + a break of only 10 min + 120 min of manual handling).

The lifting activities are performed at different heights (of the hands) at origin and destination and different horizontal distances. A significant control is present for all lifting actions. There is minimal lift asymmetry for all lifts (i.e., all objects are lifted in front of the body resulting in an asymmetry multiplier = 1.0), and the hand-to-object coupling is poor for all lifts (i.e., coupling multiplier = 0.9). Data regarding the geometries, by weight characteristic, are shown in Table 8.28.

The same data, if the appropriate software is used (see software ERGOepm_VLI_AP), appear in the worksheet as reported in Figure 8.7. Only for this example, data regarding the weights are considered for their exact value and not (see Table 8.2) as mean values of 1 kg intervals.

TABLE 8.26
Exercise 7: Sequence and Duration of Lifting Task, Light Work, and Breaks for the Case Study in an 8 h Shift

Sequence and Duration of Task during the Shift (in min)	MANUAL LIFTING TASK	Other Light Task or Break	MANUAL LIFTING TASK	Other Light Task or Break	MANUAL LIFTING TASK	Other Light Task or Break	MANUAL LIFTING TASK	Other Light Task or Break
Minutes	120	10	120	60	120	10	60	40
Shift starts/ends at	Start: 08:00							End: 17:00
Notes		Break		Lunch		Break		

TABLE 8.27
Exercise 7: Type of Weights and Number of Boxes Lifted by the Worker during an 8 h Shift and Consequent Lifting Frequency per Type of Weight

Number of Boxes	Weight	Frequency of Lifts per Minute
494	6 kg	1.18
1,235	8 kg	2.94
123	13 kg	0.29
1,852	All boxes	4.41

In this scenario, it is not advisable to use the traditional multitask lifting index (CLI) approach, since there would be up to 50 different individual FILI values (or more than 100 if one considers both origin and destination). Also, the mean frequency of each type of lift would be very low. Since the traditional CLI approach cannot work, the proposed VLI approach, using weight and geometries simplifications, should be used to assess the task.

In the presented scenario we have only three weight categories (6, 8, and 13 kg); each of them could have two simplified variants for height of hands (good, bad) at origin/destination; in turn each of them could have one, two, or three simplified variants for horizontal distance (near, mid, far). Since, in this case, different horizontal distances per weight categories are clearly identified at both the origin and the

TABLE 8.28
Exercise 7: Data Regarding Load and Geometry Characteristics for Case Study

Load Characteristics		Origin		Destination		Number of Subtasks Derived (individual LI to be potentially computed)
Number	Weight	Height above Floor	Horizontal Distance	Height above Floor	Horizontal Distance	
494	6 kg	8 levels from 14 to 84 cm: one every 10 cm	35, 45, 55 cm	80 cm	30 cm	24
1,235	8 kg	4 levels from 80 to 110 cm: one every 10 cm	30 cm	8 levels from 14 to 84 cm: one every 10 cm	35, 45, 55 cm	96
123	13 kg	2 levels, at 30 and 50 cm	45 cm	80 cm	30 cm	2

ALL WEIGHT CATEGORIES: ANALYTICAL OBJECTS DISTRIBUTION IN THE DIFFERENT GEOMETRIES PER VERTICAL AND HORIZONTAL AREAS

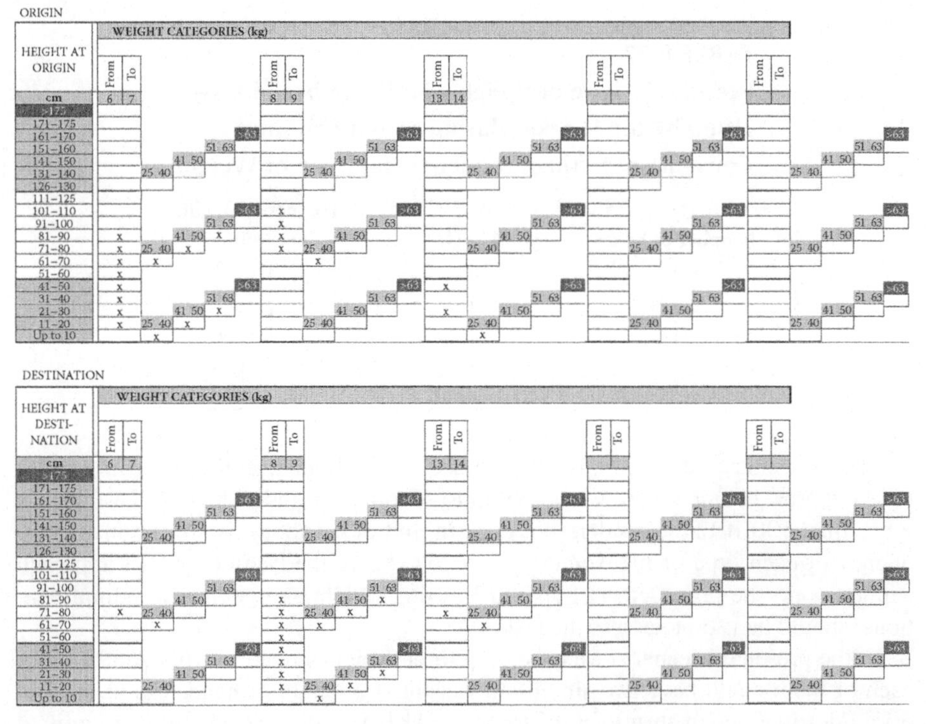

FIGURE 8.7 Exercise 7. Vertical and horizontal geometries analytically reported for the three weight categories at origin and destination.

destination, it is necessary to use the more analytical models that have been previously presented for Exercise 6.

In the following, the approaches and computations by model 1 (Section 8.4.2) and model 2 (Section 8.4.3) will be detailed with reference to Exercise 7.

8.4.2 Criteria and VLI Computation by a First Model (Exercise 7)

The first analytical model (already detailed for Exercise 6BIS) will be now used.

By using this model a total of 14 individual subtasks could be identified and corresponding FILI and STLI values can be computed; they are 6 for the first weight category (6 kg), 6 for the second weight category (8 kg), and 2 for the third weight category (13 kg).

Table 8.29 reports details regarding the 14 identified subtasks with corresponding weights, geometry multipliers, frequency multipliers, and FILI and STLI values.

Since the subtasks are more than 10, it is advisable to use the VLI concept and approach.

Subtasks and corresponding data (FILI, frequencies, and STLI) from Table 8.29 are distributed and aggregated into six LI categories.

Those six FILI categories are determined according to the distribution of the individual FILI values (in this case 14 values) using the sextile value distributions as key points for grouping (or in other terms, the values corresponding to the 16.6th, 33.3th, 50th, 66.6th, and 83.3th percentiles of the resulting FILI distribution).

The original frequencies of individual subtasks (14 in the present case) are grouped and cumulated in the six LI categories. Single (category) LI values could be consequently computed and used for reordering (from highest to lowest) the six LI categories.

Within each resulting LI category a representative FILI value is chosen: This value is the highest for the highest category (category number 1); it corresponds to the mean (central) value for the other five LI categories.

Tables 8.30 and 8.31 give details of this procedure according to the previous example.

Using these data, organised in six LI categories, it will be possible to compute the final VLI by means of the traditional CLI formula in the same way previously reported. Based on the data presented for this example, the computational data for applying the formula $VLI = STLI_1 + \Sigma\Delta LI$ are reported in Table 8.32.

Using data reported in Table 8.32, the VLI for this variable lifting task can be calculated as follows:

$$VLI = STLI_1 + \Sigma\Delta LI \tag{8.4}$$

$$STLI_1 = 1.656$$

$$\Delta FILI_2 = 0.907*[(1/0.698) - (1/0.748)]) = 0.907*(0.096) = 0.087$$

$$\Delta FILI_3 = 0.805*[(1/0.672) - (1/0.698)] = 0.805*(0.055) = 0.045$$

$$\Delta FILI_4 = 0.694*[(1/0.590) - (1/0.672)] = 0.694*(0.207) = 0.144$$

TABLE 8.29

Exercise 7, Model 1: Data Regarding the 14 Resulting Subtasks (and Corresponding FILI and STLI) by Applying the Simplification Procedures

Subtask	Weight (kg)	Vertical Height Classification and VM		Vertical Dislocation (*DM* = 1; constant)		Horizontal Distance Classification and HM		Asymmetry		Type of Grasp (*CM* = 0.9; constant)		FIRWL	FILI	Frequency (rounded)	Duration Scenario and FM (rounded)		STLI (rounded)
1	6	*L/H*	0.78	*G*	1.00	*N*	0.71	*A*	1.00	*P*	0.90	11.5	0.523	0.17	*LD*	0.850	0.62
2	6	*L/H*	0.78	*G*	1.00	*M*	0.56	*A*	1.00	*P*	0.90	9.0	0.664	0.17	*LD*	0.850	0.78
3	6	*L/H*	0.78	*G*	1.00	*F*	0.40	*A*	1.00	*P*	0.90	6.5	0.929	0.17	*LD*	0.850	1.09
4	6	*G*	1.00	*G*	1.00	*N*	0.71	*A*	1.00	*P*	0.90	14.7	0.408	0.33	*LD*	0.833	0.49
5	6	*G*	1.00	*G*	1.00	*M*	0.56	*A*	1.00	*P*	0.90	11.6	0.518	0.16	*LD*	0.850	0.61
6	6	*G*	1.00	*G*	1.00	*F*	0.40	*A*	1.00	*P*	0.90	9.3	0.725	0.16	*LD*	0.850	0.85
7	8	*L/H*	0.78	*G*	1.00	*N*	0.71	*A*	1.00	*P*	0.90	11.5	0.698	0.33	*LD*	0.833	0.84
8	8	*L/H*	0.78	*G*	1.00	*M*	0.56	*A*	1.00	*P*	0.90	9.0	0.885	0.33	*LD*	0.833	1.06
8	8	*L/H*	0.78	*G*	1.00	*F*	0.40	*A*	1.00	*P*	0.90	6.5	1.239	0.33	*LD*	0.833	1.49
10	8	*G*	1.00	*G*	1.00	*N*	0.71	*A*	1.00	*P*	0.90	14.7	0.544	0.98	*LD*	0.752	0.72
11	8	*G*	1.00	*G*	1.00	*M*	0.56	*A*	1.00	*P*	0.90	11.6	0.690	0.49	*LD*	0.811	0.85
12	8	*G*	1.00	*G*	1.00	*F*	0.40	*A*	1.00	*P*	0.90	8.3	0.966	0.49	*LD*	0.811	1.19
13	13	*L/H*	0.78	*G*	1.00	*N*	0.71	*A*	1.00	*P*	0.90	11.5	1.134	0.20	*LD*	0.850	1.33
14	13	*G*	1.00	*G*	1.00	*N*	0.71	*A*	1.00	*P*	0.90	14.7	0.885	0.10	*LD*	0.850	1.04

Note: *L/H* = low or high, *G* = good, *N* = near, *M* = medium, *F* = far, *A* = absent, *P* = poor, *LD* = long duration.

TABLE 8.30
Exercise 7, Model 1: Identification of Key Points by the Sextile Approach Using the FILI Data Distribution from Table 8.29

	1 Key Point 16.66th Percentile	2 Key Points 33.33th Percentile	3 Key Points 50th Percentile or Median	4 Key Points 66.66th Percentile	5 Key Points 83.33th Percentile	6th Category
Key value	0.527	0.672	0.711	0.885	0.960	($FILI_{max}$) 1.239
FILI category range	0.408–0.526	0.527–0.671	0.672–0.710	0.711–0.884	0.855–0.959	0.960–1.239

$$\Delta FILI_5 = 0.604*[(1/0.475) - (1/0.590)] = 0.604*(0.410) = 0.248$$

$$\Delta FILI_6 = 0.483*[(1/0.409) - (1/0.475)] = 0.483*(0.340) = 0.164$$

$$VLI = STLI_1 + \Delta FILI_2 + \Delta FILI_3 + \Delta FILI_4 + \Delta FILI_5 + \Delta FILI_6$$

$$VLI = 1.656 + 0.087 + 0.045 + 0.144 + 0.248 + 0.164 = 2.34$$

The final VLI value for the present example is 2.34 (using a reference mass of 23 kg). If the reference mass is 25 kg (as suggested in ISO and CEN standards) the final VLI value is 2.16.

8.4.3 Criteria and VLI Computation by a Second Model (Exercise 7)

The second analytical model (already detailed in Section 8.3.3 for Exercise 6) will now be used.

By using this model, considering that a significant control is present for all the lifts, the worst-case combination of factors at the origin and destination of the lifts will be used for the possible combinations of factors.

Consequently, a total of 13 individual subtasks could be identified and corresponding FILI and STLI values can be computed.

All combinations (BN, BM, BF, GN, GM, GF), with the same prevalence, are attributed both for the first weight category (6 kg) and for the second weight category (8 kg). The third weight category (13 kg) presents one only possible combination (BN = worst case).

Table 8.33 reports details regarding the 13 identified subtasks with corresponding weights, geometry multipliers, frequency multipliers, and FILI and STLI values.

Please note the similarity of values in Tables 8.29 and 8.33. The latter differs from the previous for the number of subtasks (13), the frequencies estimated for each subtask (by the worst-case combination rule), and the FILI/STLI values for the last subtask that represents the higher weight category.

TABLE 8.31
Exercise 7, Model 1: Relevant Values for Each FILI Category Using the Key Points from Table 8.30 and the Consequent Cumulated Frequencies Derived from Table 8.29

Category Data	FILI CAT (<16.66th)	FILI CAT (16.66th–33.33th)	FILI CAT (33.33th–50th)	FILI CAT (50th–66.66th)	FILI CAT (66.66th–83.33th)	FILI CAT (>83.33th)
Range of FILI values	0.408–0.526	0.527–0.671	0.672–0.710	0.711–0.884	0.855–0.959	0.960–1.239
Representative category FILI value	0.483	0.604	0.694	0.805	0.907	1.239
Number of subtasks in each category	3	2	2	2	2	3
Cumulative frequency (lpm) within the category	0.66	1.15	0.82	0.26	0.50	1.02
FM values (long duration)	0.791	0.735	0.772	0.842	0.810	0.748
STLI (category) value	0.611	0.822	0.899	0.956	1.12	1.656
Order by STLI value	6	5	4	3	2	1

TABLE 8.32
Exercise 7, Model 1: Relevant Data for Computing Final VLI Derived from Table 8.30 for the Case Study

Connotation of Cumulative Frequencies By STLI Order	Cumulative Frequencies of Categories (lpm)	Corresponding FM(long duration)	Partial Value $[(1/FM_j) - (1/FM_{j-1})]$	FILI	$STLI_1$ and $\Delta FILI_j$
FM_1	1.02	0.748		1.239	**1.656**
$FM_{1,2}$	1.52	0.698	0.096	0.907	**0.087**
$FM_{1,2,3}$	1.78	0.672	0.055	0.805	**0.045**
$FM_{1,2,3,4}$	2.60	0.590	0.207	0.694	**0.144**
$FM_{1,2,3,4,5}$	3.75	0.475	0.410	0.604	**0.248**
$FM_{1,2,3,4,5,6}$	4.41	0.409	0.340	0.483	**0.164**

Here too, as in the previous model, subtasks and corresponding data (FILI, frequencies, and STLI) from Table 8.33 should be aggregated into six LI categories.

The six LI categories may be derived by the sextile approach or, as in the present example, by using LI category limits obtained dividing the difference (FILI Max – FILI Min) in six equal parts.

The original frequencies of individual subtasks (13 in the present case) are grouped and cumulated in the six LI categories. Single (category) LI values could be consequently computed and used for reordering (from highest to lowest) the six LI categories.

Within each resulting LI category a representative FILI value is chosen: This value is the highest for the highest category (category number 1); it corresponds to the mean (central) value for all of the other five FILI categories.

Table 8.34 gives details of this procedure according to the previous example.

Using data from Table 8.34, organised in six LI categories, it will be possible to compute the final VLI by means of the traditional CLI formula.

Based on the data presented for this example and the application of the second computational model, the data for applying the formula $VLI = STLI_1 + \Sigma\Delta LI$ are reported in Table 8.35.

Using data reported in Table 8.35, obtained applying the second computational model, the VLI for Exercise 7 can be calculated as follows:

$$VLI = STLI_1 + \Sigma\Delta LI \tag{8.4}$$

$$STLI_1 = 1.595$$

$$\Delta FILI_2 = 0.966*[(1/0.723) - (1/0.776)]) = 0.966*(0.095) = 0.092$$

$$\Delta FILI_3 = 0.907*[(1/0.654) - (1/0.723)] = 0.907*(0.146) = 0.132$$

$$\Delta FILI_4 = 0.704*[(1/0.536) - (1/0.654)] = 0.704*(0.337) = 0.237$$

TABLE 8.33
Exercise 7, Model 2: Data Regarding the 13 Resulting Subtasks (and Corresponding FILI and STLI) by Applying the Simplification Procedures

Subtask	Weight (kg)	Vertical Height Classification and VM		Vertical Dislocation (*DM* = 1; constant)		Horizontal Distance Classification and HM		Asymmetry		Type of Grasp (*CM* = 0.9; constant)		FIRWL	FILI	Frequency (rounded)	Duration Scenario and FM (rounded)		STLI (rounded)
1	6	*L/H*	0.78	*G*	1.00	*N*	0.71	*A*	1.00	*P*	0.90	11.5	0.523	0.20	*LD*	0.850	0.62
2	6	*L/H*	0.78	*G*	1.00	*M*	0.56	*A*	1.00	*P*	0.90	9.0	0.664	0.20	*LD*	0.850	0.78
3	6	*L/H*	0.78	*G*	1.00	*F*	0.40	*A*	1.00	*P*	0.90	6.5	0.929	0.20	*LD*	0.850	1.09
4	6	*G*	1.00	*G*	1.00	*N*	0.71	*A*	1.00	*P*	0.90	14.7	0.408	0.20	*LD*	0.850	0.48
5	6	*G*	1.00	*G*	1.00	*M*	0.56	*A*	1.00	*P*	0.90	11.6	0.518	0.20	*LD*	0.850	0.61
6	6	*G*	1.00	*G*	1.00	*F*	0.40	*A*	1.00	*P*	0.90	9.3	0.725	0.20	*LD*	0.850	0.85
7	8	*L/H*	0.78	*G*	1.00	*N*	0.71	*A*	1.00	*P*	0.90	11.5	0.698	0.49	*LD*	0.811	0.86
8	8	*L/H*	0.78	*G*	1.00	*M*	0.56	*A*	1.00	*P*	0.90	9.0	0.885	0.49	*LD*	0.811	1.09
8	8	*L/H*	0.78	*G*	1.00	*F*	0.40	*A*	1.00	*P*	0.90	6.5	1.239	0.49	*LD*	0.811	1.53
10	8	*G*	1.00	*G*	1.00	*N*	0.71	*A*	1.00	*P*	0.90	14.7	0.544	0.49	*LD*	0.811	0.67
11	8	*G*	1.00	*G*	1.00	*M*	0.56	*A*	1.00	*P*	0.90	11.6	0.690	0.49	*LD*	0.811	0.85
12	8	*G*	1.00	*G*	1.00	*F*	0.40	*A*	1.00	*P*	0.90	8.3	0.966	0.49	*LD*	0.811	1.19
13	13	*L/H*	0.78	*G*	1.00	*N*	0.71	*A*	1.00	*P*	0.90	11.5	1.134	0.29	*LD*	0.838	1.35

TABLE 8.34
Exercise 7, Model 2: Identification of Key Points by the (Max – Min)/6 Approach and Relevant Values for Each LI Category Using the FILI Data Distribution from Table 8.33

Category Data	(FILI Max – FILI Min)/6 = 1.239 – 0.408/6 = 0.138					
Range of FILI values	0.408–0.546	0.547–0.684	0.685–0.822	0.823–0.961	0.962–1.099	1.100–1.239
Representative category FILI value	0.498	0.564	0.704	0.907	0.966	1.239
Number of subtasks in each category	4	1	3	2	1	2
Cumulative frequency (lpm) within the category	1.09	0.2	1.18	0.69	0.49	0.78
FM values (long duration)	0.741	0.350	0.732	0.787	0.811	0.776
STLI (category) value	0.673	0.781	0.962	1.152	1.191	1.595
Order by STLI value	6	5	4	3	2	1

TABLE 8.35
Exercise 7, Model 2: Relevant Data for Computing Final VLI Derived from Table 8.34 for the Case Study

Connotation of Cumulative Frequencies by STLI Order	Cumulative Frequencies of Categories (lpm)	Corresponding FM(long duration)	Partial Value $[(1/FM_j) - (1/FM_{j-1})]$	FILI	$STLI_1$ and $\Delta FILI_j$
FM_1	0.78	0.776		1.239	**1.595**
$FM_{1,2}$	1.27	0.723	0.095	0.966	**0.092**
$FM_{1,2,3}$	1.96	0.654	0.146	0.907	**0.132**
$FM_{1,2,3,4}$	3.14	0.536	0.337	0.704	**0.237**
$FM_{1,2,3,4,5}$	3.34	0.516	0.072	0.664	**0.048**
$FM_{1,2,3,4,5,6}$	4.41	0.409	0.519	0.498	**0.259**

$$\Delta FILI_5 = 0.664*[(1/0.516) - (1/0.536)]) = 0.664*(0.072) = 0.048$$

$$\Delta FILI_6 = 0.498*[(1/0.409) - (1/0.516)] = 0.498*(0.519) = 0.259$$

$$VLI = STLI_1 + \Delta FILI_2 + \Delta FILI_3 + \Delta FILI_4 + \Delta FILI_5 + \Delta FILI_6$$

$$VLI = 1.595 + 0.092 + 0.132 + 0.237 + 0.048 + 0.259 = 2.36$$

The final VLI value for the present example, computed by a second model, is 2.36 (using a reference mass of 23 kg).

If the reference mass is 25 kg (as suggested in ISO and CEN standards) the final VLI value is 2.17.

8.4.4 Final Remarks Concerning Exercise 7

Exercise 7 concerns a lifting task at an assembly line.

In this example different vertical heights and horizontal distances per each weight category are clearly identified at both the origin and the destination.

This consideration leads to using the more analytical models that have been presented and avoiding use of simplified models that consider similar horizontal distances for all weight categories.

As already stated, these simplified models should be used only when horizontal distances are practically similar in all weight categories or it is impossible to distinguish in detail all the horizontal distances present in each of the weight categories (as presented for Exercise 5 and partially for Exercise 6).

For analysing Exercise 7, two analytical computational models (already presented for Exercise 6) were used.

They refer to the same general framework and differ only for minor details regarding the way of estimating subtasks, respective individual frequencies, and FILI, and to the procedures for the aggregation of final LI categories (sextile approach vs. (Max – Min)/6 approach).

Despite these minor differences, the two computational models, also for Exercise 7, produced very similar, practically identical, results.

The reader was given information on both models for a detailed knowledge about the correspondent computational steps; the choice regarding the use of the alternative models is thus remitted to users and to availability of supporting software with the awareness that they lead to similar results.

9 Computing the Exposure Index for Evaluating Sequential Tasks: Criteria and Procedures

A sequential task occurs when workers rotate between a series (two or more) of single or multitask (composite or variable) lifting rotation slots during a work shift. Each lifting task should be performed continuously for at least 30 min. The calculation methodology to be used is the sequential lifting index (SLI) [Waters et al., 2007]. In this case, what is measured is a series of independent, separate lifting tasks as performed sequentially during a predetermined period of time (time sequence).

Figure 9.1 (Example 1) shows three sample shifts, all featuring the same two lifting tasks, task A and task B, with the same duration during the shift, but with different rotation schedules.

If one considers that task A is more physically stressful than task B, one would assume that scenario 3—in which the two tasks are alternated (i.e., rotated) every hour—would be the least stressful.

What determines the difference in exposure level among the three scenarios is the *intrinsic task duration*, which is defined by its *longest consecutive duration*. In scenario 3, task A has an *intrinsic duration* of 60 min (Figure 9.1), the shortest one in the three alternative scenarios. In the three examples, the total duration of task A is always 240 min, corresponding to a time fraction (TF) of 50%. TF is the ratio (sometimes expressed in percentage terms) of the total duration (in minutes) of each task in the shift as related to a constant of 480 min.

Table 9.1 reports the formula to compute the SLI.

Tables 9.2 and 9.3 indicate the preliminary organisational data and the operations required to obtain the final SLI for exposure to sequential lifting tasks (considering the example of Figure 9.1).

Let's consider now another example (Exercise 8). Consider two manual lifting tasks, one more stressful (higher intrinsic LI) (A) and the other a little less stressful (B). The same table describes the main characteristics and precalculated lifting indexes (LIs) for all different duration scenarios that could occur: short, medium, and long for tasks A and B.

In order to compute the SLI, the following Tables 9.5 to 9.7 present examples (Exercise 8) of how to estimate the *intrinsic duration* of a task, *total duration*, and computation of the *time fraction* (TF) for various task rotation scenarios (Exercises 8.1 to 8.3).

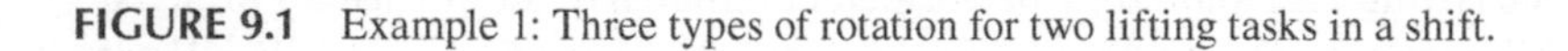

FIGURE 9.1 Example 1: Three types of rotation for two lifting tasks in a shift.

TABLE 9.1
Formula for Computing SLI

$$SLI = LI_{1\ intr} + [(LI_{1\ max} - LI_{1\ intr}) \times K]$$

where

$$K = \frac{\Sigma[(LI_{1\ max} \times TF_1) + \ldots + (LI_n \times TF_n)]}{LI_{1\ max}}$$

and

$LI_{1\ intr}$ = Lifting index for the most overloading task calculated for its intrinsic duration.
LI_{max} = Lifting index for the most overloading task calculated for the total duration of lifting activities.
LI_i = Lifting index of task i.
TF = % actual duration of the various tasks with respect to a "constant" (480 minutes).

TABLE 9.2
Organisational Characteristics of the Three Rotation Scenarios in Example 1

	Intrinsic Duration (min)	Duration Multiplier	Actual Duration (min)	Time Fraction (%)	
A	240	Long	240	50%	Scenario 1
B	180	Long	180	37.5%	
A	120	Medium	240	50%	Scenario 2
B	120	Medium	180	37.5%	
A	60	Short	240	50%	Scenario 3
B	60	Short	180	37.5%	

TABLE 9.3
Organisational Data Required to Estimate SLI

1. Determining intrinsic duration of each task to calculate intrinsic lifting index (LI_{intr}).
   The intrinsic duration of a task is determined by its longest consecutive duration in the shift (no other events taken into account).
2. Determining total duration of lifting in the shift to calculate maximum lifting index (LI_{max}):
   Use same frequency as LI_{intr} but refer to total duration scenario.
   The maximum (total) duration of a task is determined by the total duration of lifting tasks in the shift (taking into account other events such as breaks or light work not involving manual lifting).
3. Determining fractions or percentages (%) of duration of each task in relation to a constant 480 min to calculate respective TF (time fraction).

TABLE 9.4
Exercise 8: LI Precalculated for Two Tasks (A and B) Considering All the Different Duration Scenarios

Task	Weight of Load	Frequency (lifts/min)	LI (short duration)	LI (medium duration)	LI (long duration)
A	10 kg	8	1.77	3.03	5.89
B	15 kg	4	1.13	1.32	2.12

Table 9.8 shows how to compute the SLI for a first scenario (Exercise 8.1). The other scenarios (Exercise 8.2 and 8.3, reported respectively in Table 9.6 and Table 9.7) lead obviously to lower SLI values, where task A alternates with task B every 2 h (Table 9.5).

It should be noted that the mathematical formula bases its criteria on computing the lifting index for the heaviest task with respect to both its intrinsic and total duration. The same procedure is extended to other tasks, also ranked from the highest to the lowest intrinsic exposure level.

In other words, the final SLI value is always somewhere in the range of values between:

- The LI of the most physically stressful task, with its frequency, for the intrinsic duration of the task
- The LI of the same task, with the same frequency, for the total duration of the lifting tasks

This assumption has a practical relevance: Once the most stressful task has been identified, one may simply estimate the range (minimum, maximum) of SLI expected values. This estimation could be sufficient for classification purposes or to address consequent preventive measures. If not, the formula in Table 9.1 should be used.

The formula in Table 9.1 is essentially devoted to compute the value of K (see equation in Table 9.1), which identifies the true SLI value within the range, given the contribution (and sequence) of the other tasks included in the analysis.

TABLE 9.5
Exercise 8.1: Determining LI_{max}, LI_{intr}, and TF for Task A, Which Alternates with Task B Every 2 h throughout the Entire Shift

1. Distribution of Tasks in the Shift								
A	A	B	B	Lunch Break	A	A	B	B
60 min	60 min	60 min	60 min		60 min	60 min	60 min	60 min

2. Determining the Intrinsic Duration of Each Task				
	Longest Consecutive Period	Intrinsic Duration Factor	Actual Duration in the Shift (min)	TF (%)
A	120	Medium	240	50%
B	120	Medium	240	50%

3. Determining Total Duration Scenario: Long ($A + A + B + B = 240$ min)

4. Determining LI_{max}, LI_{intr}, and TF (%)				
		LI_{max}	LI_{intr}	TF (%)
1 (worst)	A	5.89	3.03	50
2	B	2.12	1.32	50

Moreover consider that, with respect to the original proposal [Waters et al., 2007], the constant for computing the time fraction (TF) was changed from 240 to 480 (min). This slight change was introduced for better representing the amount of time spent in different lifting tasks during an ordinary shift also considering possible differences during various parts of the shift (i.e., morning, afternoon). For better explaining, consider the previous example (Exercise 8.1) as reported (with computations) in Table 9.8.

The sequence of lifting tasks in the job during the shift is reported in Table 9.9.

Consider now the same job and lifting tasks alternating in the same way, but only in the morning (light work in the afternoon) as reported in Table 9.10.

Using the 240 min constant in computing K, the SLI results of these two scenarios (Tables 9.9 and 9.10) would be the same. If referred, for instance, to data in Table 9.8, the SLI for both scenarios will be 4.97. Using the 480 min constant the results for the first scenario (Table 9.9) will always be SLI = 4.97, but for the second (Table 9.10) will become SLI = 4.0: This example shows how, maintaining the fundamentals of the RNLE, especially for what concerns frequency and duration (i.e., >2 h = long duration), the choice of 480 min as a constant, for computing TF and K in the SLI formula, better reflects differences in the amount of time spent in different lifting tasks (and light work) during the whole shift.

TABLE 9.6
Exercise 8.2: Determining LI_{max}, LI_{intr}, and TF for Task A (2 h), Followed by Light Work (1 h), Followed by Task B (1 h)

1. Distribution of Tasks in the Shift								
A	A	Light Work	B	Lunch Break	A	A	Light Work	B
60 min	60 min	60 min	60 min		60 min	60 min	60 min	60 min

2. Determining the Intrinsic Duration of Each Task				
	Longest Consecutive Period	Intrinsic Duration Factor	Actual Duration in the Shift (min)	TF (%)
A	120	Medium	240	50%
B	60	Short	120	25%
3. Determining Total Duration Scenario: Moderate (A + A = 120 min)				

4. Determining LI_{max}, LI_{intr}, and TF (%)				
		LI_{max}	LI_{intr}	TF (%)
1 (worst)	A	3.02	3.03	50%
2	B	1.32	1.13	25%

Going back to the formula, it is clear that considering all its determinants is rather complex, especially where there are combinations of alternating composite or variable tasks in a sequence; therefore specific software needs to be used. Such is the complexity and quantity of data to be entered that unfortunately it has thus far not been possible to create a unique tool for computing this SLI by a common spreadsheet, unlike for other lifting indexes.

Therefore the authors decided to develop a set of dedicated software (always using a common spreadsheet) to be used in sequence, trying to maintain user-friendly characteristics. The software can be found and downloaded freely at www.cpmresearch.org. To obtain the final calculation of SLI it is necessary to use dedicated software divided into three sections:

- ERGOepm_SLI: For collecting the initial organisational data and for the final computation of the SLI.
- ERGOepm_SLI_LI&CLI: For calculating the LI for single task or composite task with less than 10 subtasks.
- ERGOepm_SLI_CLI&VLI: For calculating the LI for composite task with more than 10 subtasks or variable tasks.

TABLE 9.7

Exercise 8.3: Determining LI_{max}, LI_{intr}, and TF for Task A (1 h), Followed by "Light" Work (1 h), Followed by Task B (1 h)

1. Distribution of Tasks in the Shift									
A	Light Work	B	Light Work	Lunch Break	A	Light Work	B	Light Work	
60 min	60 min	60 min	60 min		60 min	60 min	60 min	60 min	

2. Determining the Intrinsic Duration of Each Task				
	Longest Consecutive Period	Intrinsic Duration Factor	Actual Duration in the Shift (min)	TF (%)
A	60	Short	120	25%
B	60	Short	120	25%

3. Determining Total Duration Scenario: Short (A = 60 min)

4. Determining LI_{max}, LI_{intr}, and TF (%)				
		LI_{max}	LI_{intr}	TF (%)
1 (worst)	A	1.77	1.77	25%
2	B	1.13	1.13	25%

Rather than theoretically explaining how the formula is applied, Exercise 9 shows what data need to be collected and entered into the software: Once the correct information is entered, the final lifting index (SLI) is then generated.

The first information that needs to be collected concerns (Figure 9.2):

- The types of tasks performed during the shift (i.e., single or composite or variable): See ERGOepm_SLI,sheet 1.
- The distribution and duration of the tasks during the shift (i.e., the shift log for sequential tasks): See ERGOepm_SLI,sheet 1.

The next information to be acquired concerns the specific characteristics of each individual task (Figures 9.3 to 9.6) in terms of:

- Type, weight, and number of objects to be lifted
- Estimated lifting frequency of task
- Analysis of load geometries at origin and destination
- Twisting of the torso (asymmetry)
- Type of coupling

TABLE 9.8
Exercise 8.1: Example of Computing SLI for the Rotation Pattern Reported in Table 9.4

1. Distribution of Tasks in the Shift								
A	A	B	B	Lunch Break	A	A	B	B
60 min	60 min	60 min	60 min	60 min	60 min	60 min	60 min	60 min

Task Order		LI_{max}	LI_{intr}
First	A	5.89	3.03
Second	B	2.12	1.32

TF of task A = 240/480 = 0.5 (50%)
TF of task B = 240/480 = 0.5 (50%)

$$K = \frac{[(LI_{1\ max} \times FT_1) + \ldots + (LI_n \times TF_n)]}{LI_{1\ max}}$$

$$K = \frac{[(5.89 \times 0.5) + (2.12 \times 0.5)]}{5.89} = 0.68$$

$$SLI = LI_1 + [(LI_{1\ max} - LI_1) \times K]$$

$$SLI = 3.03 + [(5.89 - 3.03) \times 0.68]$$

$$SLI = 4.97$$

Traditional computation models (Chapters 4 to 6) are used to evaluate the intrinsic characteristics of single or composite tasks featuring only a few subtasks: In this case use the model ERGOepm_SLI_LI&CLI; for composite or variable tasks with more than 10 subtasks, the new computation models presented in Chapters 7 and 8 are used (ERGOepm_SLI_CLI&VLI).

The software, developed to compute sequential tasks, provides a single guide for entering these data, and then automatically supplies the SLI exposure index.

In writing the general information about the tasks' distribution in the shift with their characteristics (ERGOepm_SLI,sheet 1), it is necessary to calculate the intrinsic lifting index of each task present in the sequential distribution. To obtain this evaluation, open the specific software for each task performed, as specified above.

Obtain the intrinsic LI for each task using the specific software model ERGOepm_SLI_LI&CLI or model ERGOepm_SLI_CLI&VLI. It is necessary to copy the results again in ERGOepm_SLI,sheet 1.

Figure 9.7 illustrates the intrinsic LI values for each task A, B, and C obtained.

Figure 9.3 shows, with reference to Exercise 9, how to estimate the total duration of the sequential tasks in the shift, although the software produces this calculation automatically.

In general it should be emphasised that when the most physically stressful task is the most highly represented (and the contribution of other tasks is negligible), the result will be very close to the one obtained for this task.

TABLE 9.9
Exercise 8.1: Task A Alternates with Task B Every 2 h throughout the Entire Shift

1. Distribution of Tasks in the Shift								
A	A	B	B	Lunch Break	A	A	B	B
60 min	60 min	60 min	60 min	60 min	60 min	60 min	60 min	60 min

TABLE 9.10
Exercise 8.1: Task A Alternates with Task B Every 2 h Only in the Morning

1. Distribution of Tasks in the Shift								
A	A	B	B	Lunch Break	Light Work	Light Work	Light Work	Light Work
60 min	60 min	60 min	60 min	60 min	60 min	60 min	60 min	60 min

In fact in Exercise 9, the final SLI value (Figure 9.8) is identical to the intrinsic value of task C.

How is this possible? When the intrinsic duration of the most physically stressful task (long, in present case) is equal to the total task duration (long, again in this case), then LI_{1max} equals LI_{1intr}: This practically cancels out K from the general SLI formula.

Hence it is obvious that in this case, the SLI value is the same as the LI_{1intr} of the most stressful task.

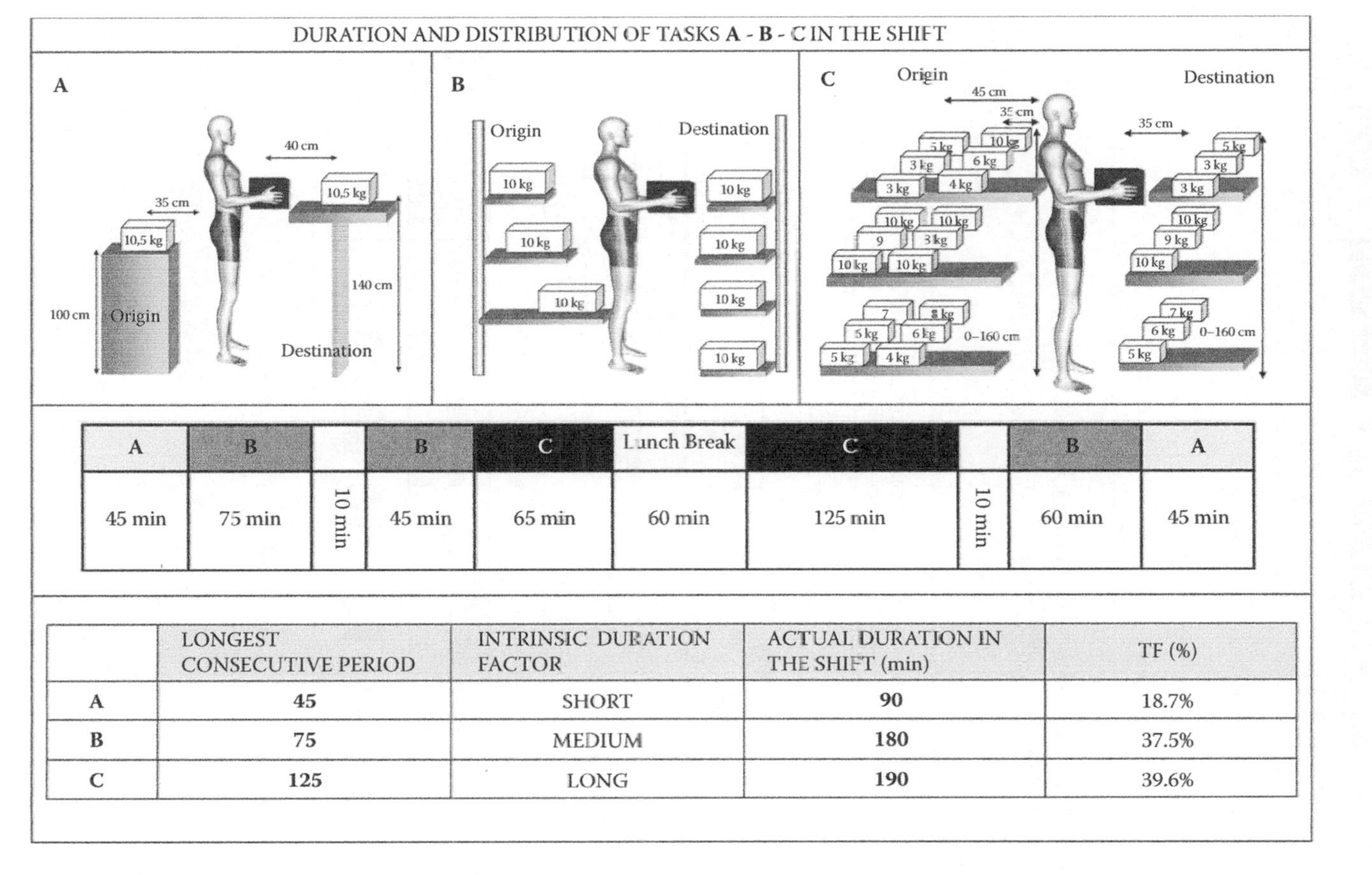

A	B		B	C	Lunch Break	C		B	A
45 min	75 min	10 min	45 min	65 min	60 min	125 min	10 min	60 min	45 min

	LONGEST CONSECUTIVE PERIOD	INTRINSIC DURATION FACTOR	ACTUAL DURATION IN THE SHIFT (min)	TF (%)
A	45	SHORT	90	18.7%
B	75	MEDIUM	180	37.5%
C	125	LONG	190	39.6%

FIGURE 9.2 Exercise 9: Sequential tasks. Organisational characteristics and structure of shift rotation.

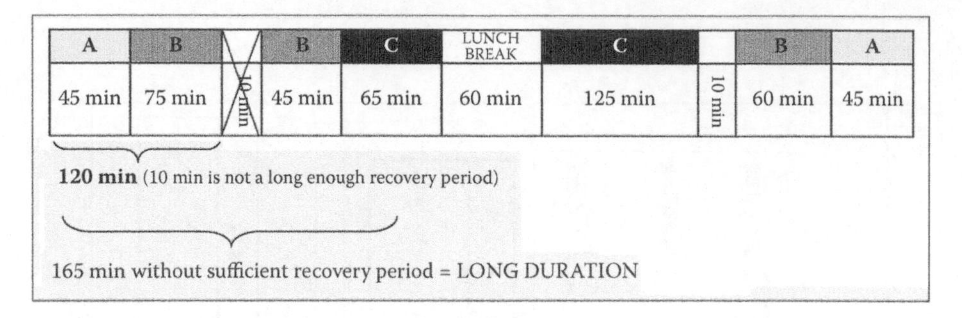

FIGURE 9.3 Exercise 9: Sequential tasks. Estimated total duration of lifting (see Chapter 4 for details on estimating duration).

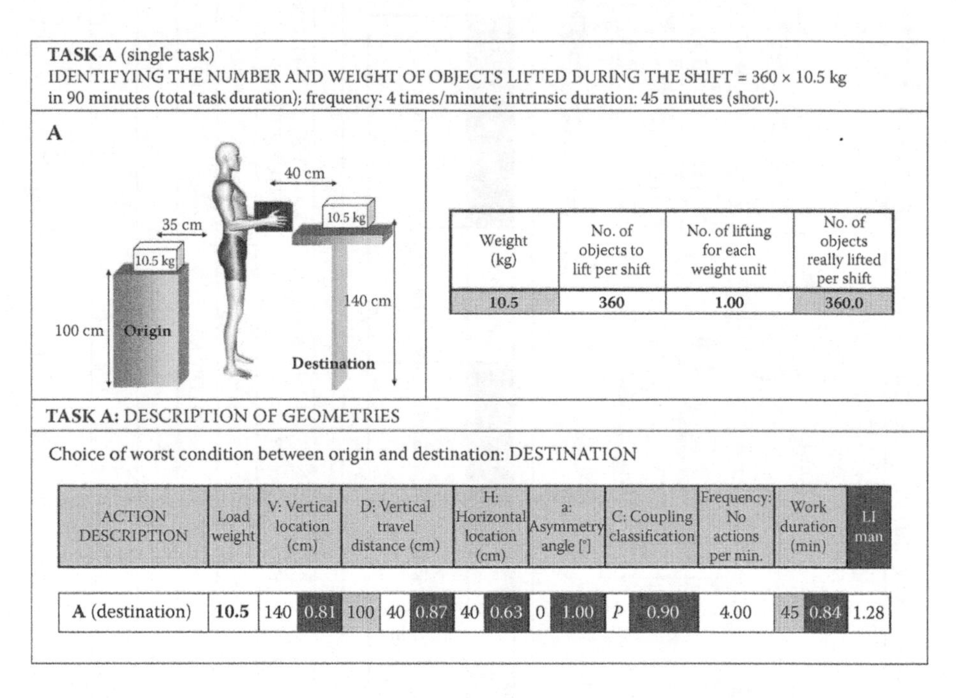

TASK A (single task)
IDENTIFYING THE NUMBER AND WEIGHT OF OBJECTS LIFTED DURING THE SHIFT = 360 × 10.5 kg in 90 minutes (total task duration); frequency: 4 times/minute; intrinsic duration: 45 minutes (short).

Weight (kg)	No. of objects to lift per shift	No. of lifting for each weight unit	No. of objects really lifted per shift
10.5	360	1.00	360.0

TASK A: DESCRIPTION OF GEOMETRIES

Choice of worst condition between origin and destination: DESTINATION

ACTION DESCRIPTION	Load weight	V: Vertical location (cm)		D: Vertical travel distance (cm)		H: Horizontal location (cm)		a: Asymmetry angle [°]		C: Coupling classification		Frequency: No actions per min.	Work duration (min)		LI man
A (destination)	**10.5**	140	0.81	100	40	0.87	40	0.63	0	1.00	*P*	0.90	4.00	45	0.84	1.28

FIGURE 9.4 Exercise 9: Single task A. Intrinsic characteristics.

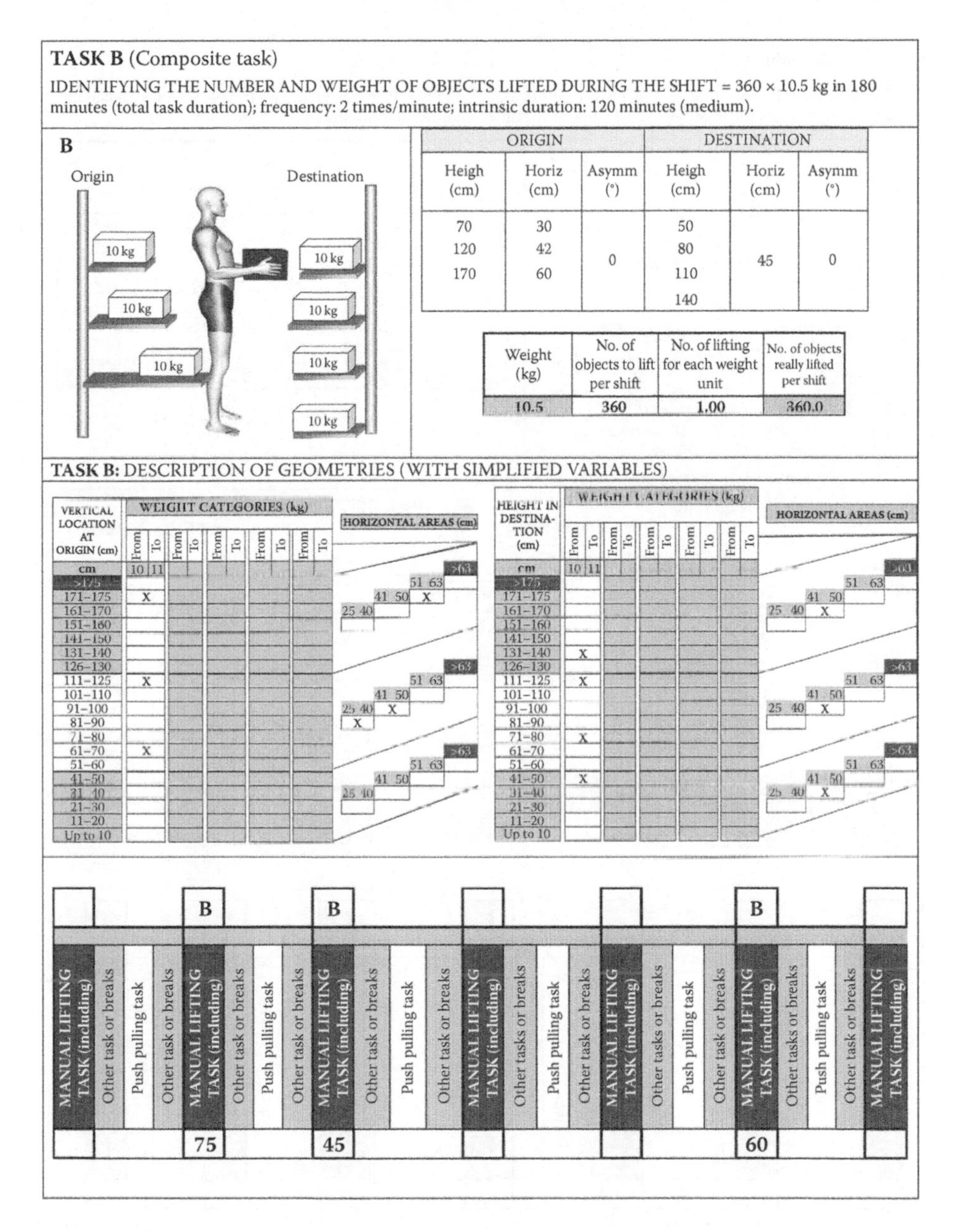

FIGURE 9.5 Exercise 9: Composite task B. Intrinsic characteristics.

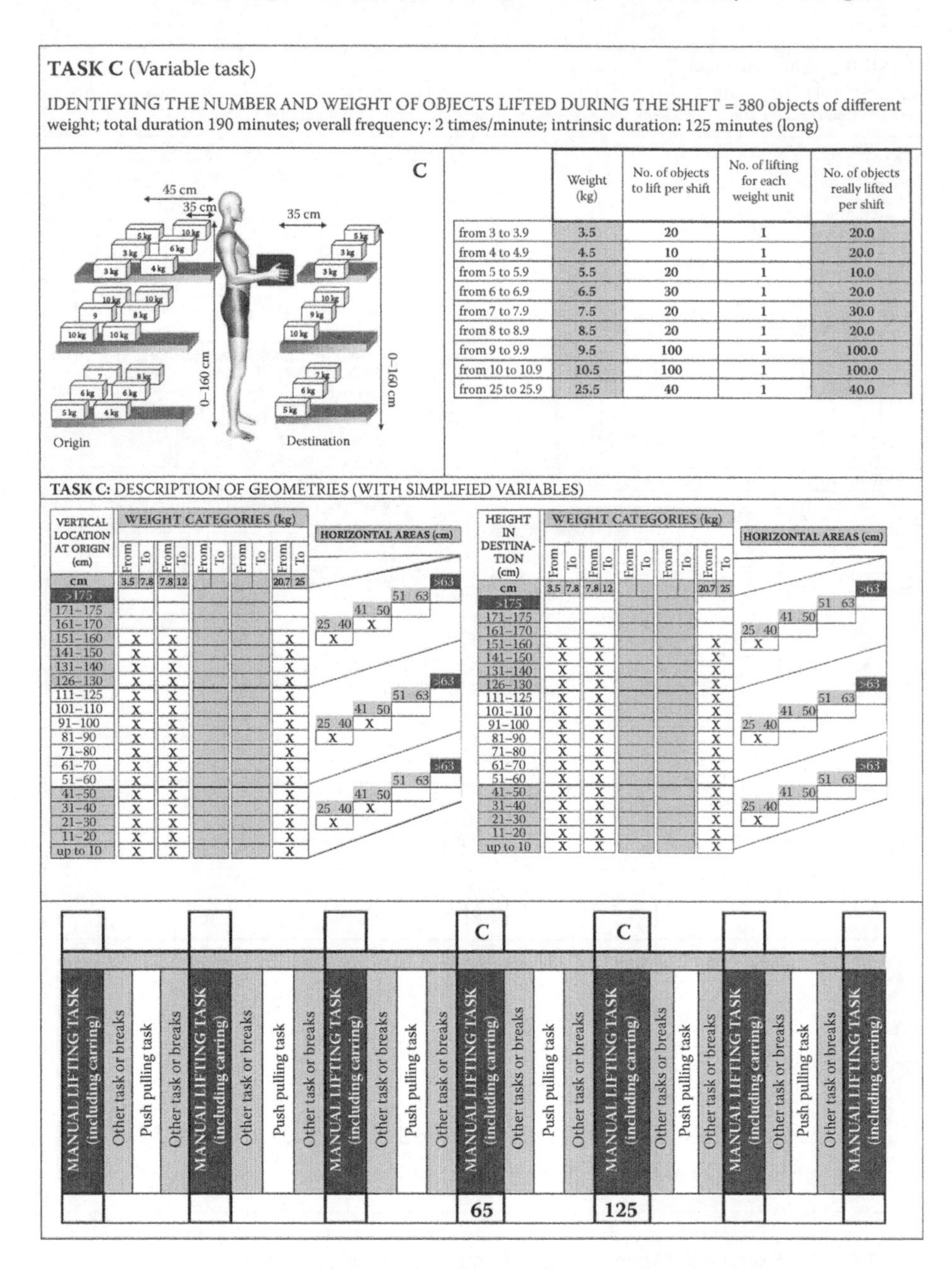

TASK C (Variable task)

IDENTIFYING THE NUMBER AND WEIGHT OF OBJECTS LIFTED DURING THE SHIFT = 380 objects of different weight; total duration 190 minutes; overall frequency: 2 times/minute; intrinsic duration: 125 minutes (long)

	Weight (kg)	No. of objects to lift per shift	No. of lifting for each weight unit	No. of objects really lifted per shift
from 3 to 3.9	3.5	20	1	20.0
from 4 to 4.9	4.5	10	1	20.0
from 5 to 5.9	5.5	20	1	10.0
from 6 to 6.9	6.5	30	1	20.0
from 7 to 7.9	7.5	20	1	30.0
from 8 to 8.9	8.5	20	1	20.0
from 9 to 9.9	9.5	100	1	100.0
from 10 to 10.9	10.5	100	1	100.0
from 25 to 25.9	25.5	40	1	40.0

TASK C: DESCRIPTION OF GEOMETRIES (WITH SIMPLIFIED VARIABLES)

FIGURE 9.6 Exercise 9: Variable task C. Intrinsic characteristics.

TASK A (single task): evaluated with the revised NIOSH Equation

ACTION DESCRIPTION	Load weight	V: Vertical location (cm)		D: Vertical travel distance (cm)		H: Horizontal location (cm)		a: Asymmetry angle [°]		C: Coupling classification		Frequency: No actions per min.	Work duration (min)		LI man
A (destination)	10.5	140	0.81	100	40	0.87	40	0.63	0	1.00	P	0.90	4.00	45	0.84	1.28

TASK B (composite task): evaluated with the new simplified model for CLI (more than 10 subtasks) or VLI

European Standard (EN): 1005-2; ISO Standard: 11228-1

25	**Men** (18–45 years old)	**1.73**
20	**Women** (18–45 years old)	**2.16**
20	**Men** (<18 o >45 years old)	**2.16**
15	**Women** (<18 o >45 years old)	**2.88**

Original NIOSH Lifting equation

23	**NIOSH original**	**1.88**

TASK C (variable task): evaluated with the new simplified model for VLI

European Standard (EN): 1005-2; ISO Standard: 11228-1

25	**Men** (18–45 years old)	**3.29**
20	**Women** (18–45 years old)	**4.11**
20	**Men** (<18 o >45 years old)	**4.11**
15	**Women** (<18 o >45 years old)	**5.48**

Original NIOSH Lifting equation

23	**NIOSH original**	**3.58**

FIGURE 9.7 Exercise 9: Sequential tasks. Intrinsic LI values for tasks A, B, and C.

	Intrinsic duration (min)	Actual duration in the shift (min)	TF (%)	Intrinsic LI	Max LI
A	45 (SHORT)	90	18.7%	1.28	2.38
B	75 (MEDIUM)	180	37.5%	1.73	2.11
C	125 (LONG)	190	39.6%	3.29	3.29
SEQUENTIAL LIFTING INDEX (SLI):				3.29	

FIGURE 9.8 Exercise 9: Sequential tasks. Final SLI.

10 Mapping Risk Associated with Manual Load Lifting

It can be very helpful to map specific risk in working contexts where there are numerous jobs and workstations requiring manual lifting of loads and especially if there are workers who already suffer from work-related back problems (see Chapter 11).

The tools for collecting the information needed for risk mapping have been illustrated in earlier chapters.

A mapping program (Excel) has been designed to process data regarding load lifting and other types of manual handling tasks such as carrying, pulling, and pushing. The results may be broken down by line, operating area, department, company, or gender.

The risk mapping procedures and explanations on how to use the program are provided in this chapter.

The mapping software ERGOepm_MMHmap is available as a free download from the website www.epmresearch.org.

Before starting risk mapping, it is essential to briefly describe the way the work is organised and then identify the workstations that feature manual load lifting tasks, using the criteria explained previously to define the *key enters* and *quick assessment*.

In the early stages of the intervention, the quick assessment is useful for rapidly identifying acceptable risk workstations (green code), critical conditions (critical code) requiring immediate attention, and those for which the classic (more detailed) assessment must be performed.

The risk mapping program, to be compiled for each individual department or area, consists of several sections:

- Section 1: Contains the results of the risk assessments (sheet: MAP of the software).
- Section 2: Depicts the results of the mapping procedure in tables and graph (sheets: SUMMARY and GRAPHS of the software).

10.1 SECTION 1

This section contains six blocks, each assigned to a single line, area, or several identical workstations.

Each block contains the findings of the risk assessment for each individual task single or composite or variable (entered on one horizontal row): At the bottom of each block there is a table indicating the assessment results in terms of risk codes (percentage of workstations classified with an exposure level of acceptable, borderline, present-low, present-moderate, or present-high).

The map only describes workstations involving manual load handling (lifting or carrying or pulling/pushing) broken down according to the criteria defined by the key enters, whether or not the workstations have been described as at risk.

The work performed must be assessed without initially taking into account job rotations (sequential task). Therefore the handling task is analysed as if the worker performs the same kinds of task (single or composite or variable) for the entire shift. Remember that sequential tasks are in fact a job comprised of several tasks in rotation: The outcome of the evaluation is thus attributed not to the individual task but to the person(s) performing the same job.

Figures 10.1 to 10.3 list the information that has to be reported, in the same order as it appears in the software. Information regarding Figure 10.3 derives from procedures not reported in this manual; the indexes addressed in Figure 10.3 are substantially ratios between an actual load (carrying) or force (push/pull) and corresponding reference data suggested in ISO 11228-1 (carrying) and ISO 11228-2 (whole-body pushing/pulling; method 1), respectively.

However, not all of this information is essential for computing risk indexes. The map can be used to store analytical data, especially with a view to future interventions to redesign the way that the work is organised.

10.2 SECTION 2

The latter part of this section refers to the organisational details that serve to weight the exposure values obtained so far:

- Number of identical tasks present with the same exposure level (to manual handling)
- Number of daily shifts over which the task is distributed
- Total number of workers assigned to task(s), broken down by gender

Figure 10.4 describes the information required to obtain these weightings.

As mentioned, it is often essential to assess risk exposure separately for males and females. To interpret the results of occupational health surveillance correctly, it is in fact useful to have exposure levels broken down by gender.

Figure 10.5 shows the final risk assessment graphs for lifting tasks performed in one department or area. Risk category distributions are shown across the whole group and separately for males and females.

Similar graphs may be produced automatically for carrying, pulling, and pushing (Figure 10.6).

The mapping spreadsheets, available as a free download from www.epmresearch.org, can be easily and rapidly compiled to analyse each workstation as described in previous chapters.

Once correctly compiled, all the information to be copied online into the mapping software presented here will appear automatically and in the correct order, on the synthesis page.

TASKS	NET DURATION OF MANUAL LIFTING (LOADS EQUAL OR MORE THAN 3 kg)	KINDS OF LIFT: SINGLE = S; COMPOSITE = C; VARIABLE = V	N. OF OBJECTS LIFTED (per person)	LIFTING FREQUENCY (lifts/minute)	FREQUENCY/DURATION MULTIPLIER	Weight Categories					% in Weight Categories					TOTAL MASS (kg) LIFTED PER PERSON
						FIRST	SECOND	THIRD	FOURTH	FIFTH	FIRST	SECOND	THIRD	FOURTH	FIFTH	
A	330	V	900	2.73	0.58	3.8	9.5	16.5		24.3	89%	7%	2%	0%	2%	2600
B	330	V	1280	3.88	0.46	3.8	9.5	16.5		24.3	63%	5%	2%	0%	31%	11815
C	330	V	240	0.73	0.78	3.5	4.5	9.5			8%	83%	8%	0%	0%	580
D	180	S	1600	8.89	0.53	5.5					100%	0%	0%	0%	0%	8800
E	330	V	1620	4.91	0.36	3.5	4.5	9.5			1%	12%	86%	0%	0%	7135

FIGURE 10.1 Mapping risk. Organisational data.

	Tasks	Task A	Task B	Task C	Task D	Task E
Relevance of the Single Risk Factors	Frequency/Duration					
	Vertical Height					
	Horizontal Distance					
	Asymmetry					
	Presence of Loads More Than 25 kg					
	Presence of Loads More Than 20 kg	X	X			
	Presence of Loads More Than 15 kg	X	X			
Lifting Index	Male (18-45 years old)	2.57	3.47	1.07	0.95	Critical condition
	Female (18-45 years old) or Male (<18 or >45 years old)	3.21	4.34	1.34	1.18	Critical condition
	Female (<18 or >45 years old)	4.28	5.78	1.79	1.58	Critical condition
	NIOSH Original (RNLE)	2.79	3.77	1.16	1.03	Critical condition
Critical Conditions	Critical Frequency					
	Critical Height at Origin					
	Critical Horizontal Distance at Origin					
	Critical Height at Destination					
	Critical Horizontal Distance at Destination					X
	Critical Asymmetry					

FIGURE 10.2 Mapping risk. Results of the evaluation of manual handling: The relevance of the single risk factors and the final lifting indexes for gender and age.

Task Denomination	Carrying Index 8 h: Index (total kg carried/kg recommended mass)	Carrying Index 1 h: Index (total kg carried/kg recommended mass)	Carrying Index 1 min: Index (total kg carried/kg recommended mass)	Pushing/Pulling Index Male: Index: Initial Force Pulling	Pushing/Pulling Index Male: Index: Sustained Force Pulling	Pushing/Pulling Index Male: Index: Initial Force Pushing	Pushing/Pulling Index Male: Index: Sustained Force Pushing	Pushing/Pulling Index Female: Index: Initial Force Pulling	Pushing/Pulling Index Female: Index: Sustained Force Pulling	Pushing/Pulling Index Female: Index: Initial Force Pushing	Pushing/Pulling Index Female: Index: Sustained Force Pushing	Additional Risks: Unfavourable Object Characteristics for Manual Lifting and Carrying	Additional Risks: Working Environment Unfavourable Manual Lifting and Carrying
Task A	0.10	0.10	0.10	0.30	0.30	0.10	0.10	0.30	0.30	0.10	0.10		
Task B	0.90	0.90	0.90	0.90	0.80	0.60	0.60	1.00	0.90	0.60	0.60		
Task C	1.80	1.80	1.80	1.10	1.10	1.00	1.00	1.20	1.00	1.10	1.00	X	
Task D	3.00	3.00	3.00	2.00	1.60	1.00	0.90	2.00	1.60	1.50	1.00		X
Task E	3.70	3.70	3.70	3.20	3.10	2.80	2.00	3.30	3.00	3.10	2.00	X	X

FIGURE 10.3 Mapping risk. Results of the evaluation of manual carrying, pushing, and pulling.

	Total Number of Identical Workstations in the Shifts and Their Gender Distribution				
			Workstation Number for Gender		
Task Denomination	Kinds of Shift	Number of Workplaces Identical	Total	Number of Females	Number of Males
Task A	2	6	12	6	6
Task B	2	6	12	6	6
Task C	2	6	12	6	6
Task D	2	6	12	6	6
Task E	2	6	12	6	6

FIGURE 10.4 Mapping risk. Total number of identical workstations in the shifts and their distribution by gender.

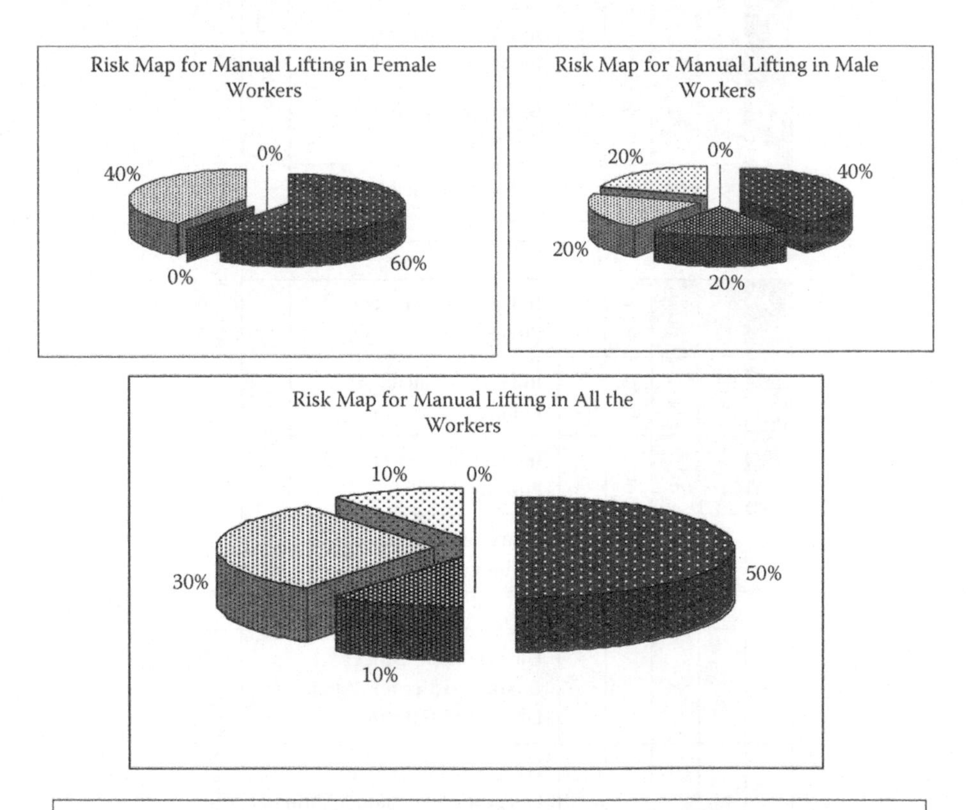

FIGURE 10.5 Mapping risk. Results of the evaluation of manual handling: The final distribution of the obtained lifting indexes for risk areas and gender (sheet GRAPH of the software).

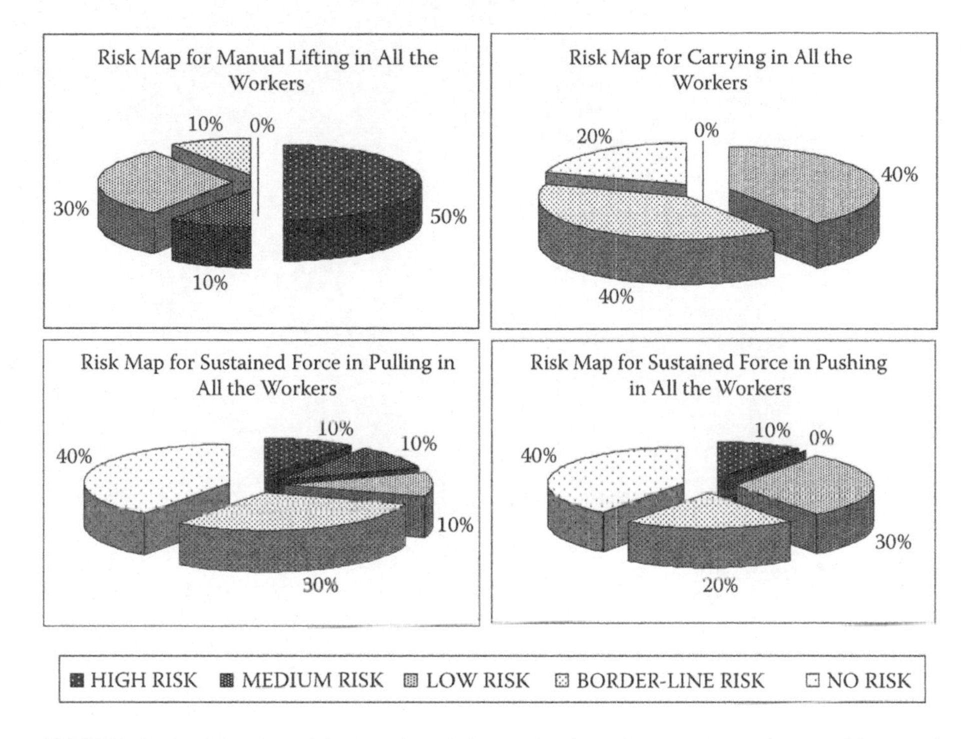

FIGURE 10.6 Mapping risk. Results of the evaluation of manual carrying, pushing, and pulling: Final distribution for risk areas (sheet GRAPH of the software).

11 Health Surveillance: Anamnestic Screening and Clinical-Functional Assessment

11.1 REGULATORY REFERENCES

Every country needs to comply with the indications and procedures laid down by its laws and regulations.

In Europe, Council Directive 89/391/EEC requires employers to ensure that workers receive health surveillance, at regular intervals, appropriate to the health and safety risks they incur at work, including exposure to manual handling activities. Health surveillance may be provided as part of a national health system. In most European countries, such health surveillance activities must be carried out by an occupational medicine physician.

Health surveillance activities should include, among others:

- A preventive medical examination to screen for conditions that might be incompatible with the job the worker has been assigned to, to ensure his or her suitability for the specific tasks (and possible risks) involved
- A periodical medical examination to monitor the worker's health

Both examinations require:

- Clinical and diagnostic tests targeting the specific risk (in our case musculoskeletal disorders due to biomechanical overload)
- A fit for job assessment

In some European countries, active health surveillance for workers exposed to manual lifting is recommended whenever a job is found to have *a lifting index higher than 1.*

11.2 AIMS OF HEALTH SURVEILLANCE

Health surveillance in general has preventive aims, and should ensure—before starting the job and then on an ongoing basis—that the relationship between the individual worker's specific health status and his or her specific working conditions is satisfactory; this relationship then needs to be assessed on a collective basis.

Next to these general aims, with regard to musculoskeletal disorders, there are also several more specific aims, such as:

- Identifying any negative health conditions at an early enough stage to prevent them from worsening
- Identifying hypersusceptible people requiring additional protective measures than those adopted for other workers
- Contributing, based on appropriate feedback, toward enhancing the accuracy of collective and individual risk assessments
- Monitoring preventive measures to ensure their continuing adequacy
- Collecting clinical data in order to compare different groups of workers and different scenarios
- Collecting data on absences caused by specific disorders, so as to estimate the cost of nonprevention

This being said, there is an obvious need to focus both on the spine and on other parts of the body. In this respect, in light of the latest literature [Silverstein et al., 2008] [Harkness et al., 2003] [Hoozemans et al., 2002], it will be necessary to broaden the scope of health surveillance to include consideration of the upper limbs, especially the shoulder.

11.3 FREQUENCY OF EXAMINATIONS

11.3.1 Preventive Examinations

Health surveillance must target individuals at the time they are hired to perform or assigned to perform jobs that involve manual load lifting.

At this time, all the workers assigned to carry out manual lifting tasks need to be examined regardless of the outcome of the risk assessment emerging from the application of *lifting indexes.*

These indexes, and the consequent indications with regard to periodical health surveillance, are based on the assumption that around 90% of the *healthy adult working population* can be protected.

At the time of hiring, the worker is examined to identify any diseases or disorders, whether occupational or nonoccupational, that could prove to be incompatible with a specific working condition, even at exposure levels regarded as relatively safe for the majority of workers.

The first examination should therefore include not only a structured patient clinical history (anamnesis) but also a clinical examination, even if the worker does not report any symptoms. Subsequent periodical examinations will require a different approach.

It should be noted that the purpose of health surveillance, at the hiring stage, should not and indeed must not be (except in rare exceptions) to select the healthiest and strongest workers to assign to physically stressful jobs, but rather to identify people who already have a disorder or disease that renders them hypersusceptible to working conditions that are acceptable to everyone and which thus calls for measures to restrict possible exposure levels [Colombini et al., 1993].

11.3.2 Periodical Examinations

As evidenced above, the aims of periodical health surveillance are somewhat different and broader in scope than routine prehiring physical examinations.

To start with, periodical surveillance should be offered to all workers exposed to manual load lifting conditions where the lifting index, at risk evaluation, was higher than 1.

For values between 0.85 and 1 (borderline risk area), clinical screening is advisable: The results of the examination could be used by the occupational medicine physician to establish if the *doubtful* risk area reveals the need to reassess the relevant risk. If the risk level remains within the doubtful area, the screening procedure should be repeated every 3 years.

The actual frequency of follow-up examinations should be decided by the occupational medicine practitioner based on the risk evaluation results and his or her knowledge of the individual and on the collective health status of the population in question. However, the physician may want to opt for a different frequency for individual workers.

Nonetheless, on the whole, it can be stated that examinations carried out every 2 or 3 years should be sufficient to monitor workers performing manual lifting tasks with a lifting index of between 1 and 3.

A commonsense approach to determining the frequency of periodical evaluations in relation to risk level might consist of the following strategy:

- Index equal to or above 3: Yearly.
- Index between 2.1 and 2.99: Every 2 years.
- Index between 1.0 and 2: Every 3 years.

The same frequency applies to younger workers and those aged over 45, provided that the risk index uses a different reference weight constant, as required by ISO 11228, part 1 [ISO, 2003].

11.4 ANAMNESTIC SCREENING PROTOCOLS

11.4.1 General Considerations

While the occupational health physician has both the right and the duty (within the framework of professional ethics, legal regulations, and the scope of health surveillance) to decide which clinical procedures to put in place, this chapter offers helpful guidelines for standardising behavioural criteria and ensuring comparable data across different environments.

Generally speaking, based on current knowledge, it is important to avoid x-rays during the screening phase; x-rays should be carried out only when deemed necessary or advisable based on the worker's preliminary history and spinal examination.

Conversely, x-rays (or CT or MR scans), alongside other instrumental, laboratory, or specialist exams (i.e., orthopaedics, rheumatology, physiatrist), should be performed where there is a definite clinical or diagnostic suspicion.

However, it should be stressed that the aforementioned procedure (screening clinical exam and diagnostic imaging only in selected cases) may once in a while generate errors in deciding preventatively whether workers are suitable for performing certain tasks.

Periodical screening initiatives may involve a three-step procedure, as follows:

- Generalised use of a structured anamnestic interview.
- Functional clinical exam of the spine, in cases positive at anamnestic interview.
- Additional specialist radiographic and instrumental exams only when required, based on anamnestic interview and clinical-functional examination of the spine. These latter examinations should be carried out when there are questions about the worker's fitness for work.

The anamnestic interview may be used to monitor the adequacy of the preventive measures adopted, or to compare clinical data among groups of workers featuring different exposure levels.

Therefore a specific health surveillance approach could be defined as envisaging the following two levels:

- *Anamnestic*, meaning an initial general level (addressing all exposed workers) to identify cases that are positive to a predefined anamnestic threshold
- *Clinical*, in other words, a second level of clinical tests addressing only workers who are positive to the anamnestic interview

In the study (and surveillance) of musculoskeletal disorders, reported symptoms are of the utmost importance since they generally have an early onset and, if detected adequately, can suggest a diagnostic suspicion and the need for the most appropriate clinical or instrumental exams.

It should be noted that in studies on large populations, even if not all workers can be visited individually, due to organisational constraints, it is nonetheless essential to somehow define the presence of significant symptoms.

Based on the foregoing considerations, there are two ways of managing the anamnestic questionnaire, for epidemiological purposes, and each way requires the data to be gathered differently:

- *Method 1*: Guided self-administered anamnestic questionnaire.
- *Method 2*: Anamnestic questionnaire administered by the occupational physician (or other suitably trained health care operator) via a direct interview, followed by a clinical-functional evaluation.

In method 1 the questionnaire is administered to groups of 15–20 exposed workers, under the supervision of a trained health care operator (health care assistant, professional nurse). The data thus gathered quickly generate an initial *rough indication* and therefore a preliminary statistic of the workers affected by more or less significant disorders (by predefining a positive threshold level) who should be called back for a clinical examination. This method can also be employed as a means of sharing information with workers: It is an excellent opportunity to explain the meaning of the disorders (what causes them and how can they be prevented), and why it is important to report them immediately to the company's health care unit without waiting for scheduled examinations.

The guided self-administered anamnestic questionnaire obviously needs to be reviewed by the company doctor to check the reliability of the responses and choose which positive cases need further clinical evaluation.

Method 2 requires that the health and safety specialist administers the anamnestic questionnaire to workers during a visit.

These approaches are designed to provide the doctor with details regarding the occurrence of *anamnestic cases* (i.e., workers affected by disorders lying above the anamnestic threshold) in special work environments, and to define clinical assessment priorities and subsequent diagnostic programs tailored to each individual worker.

11.4.2 The Model for Anamnestic Investigation

The model for anamnestic investigation includes a part focusing on the various sections of the spine and on the upper limb, including the shoulder joint.

First the respondent is asked for general information (Table 11.1): name, address, etc., company name, department, current duties, length of time in the position.

The questionnaire then goes on to define spinal disorders dating back over the last 12 months; this part is divided into three sections, cervical, dorsal, and lumbosacral, using the same rationale for defining the type and duration of the disorders.

These sections ask the worker to report painful episodes or subchronic soreness. Specific questions also relate to the location and characteristics of the disorders.

An accurate and well-defined description of symptoms helps to more accurately guide the individual diagnostic process and also permits an epidemiological comparison of data from different sources. The use of an *anamnestic threshold* and standardised data collection method for such disorders is actually what makes this comparison possible [Colombini et al., 1983, 1993] [Occhipinti et al., 1985, 1988, 1993] and enables symptoms that are not pathognomonic in terms of quality or severity to be excluded.

Table 11.2 lists the criteria for classifying *positive anamnestic cases*: A disorder is considered to be positive, i.e., over a positive threshold, when it meets the relevant conditions.

TABLE 11.1
Anamnestic Screening Protocol: Personal Details

<table>
<tr><td colspan="4">Date:</td></tr>
<tr><td>Company:</td><td colspan="2">Department:</td><td>Position:</td></tr>
<tr><td colspan="4">Name and surname:</td></tr>
<tr><td>Date of birth:</td><td>Current age:</td><td colspan="2">M/F:</td></tr>
<tr><td>Number of years in same position:</td><td colspan="3">Number of years with the company:</td></tr>
</table>

TABLE 11.2
Criteria for Defining Disorders of Various Segments of the Spine as Threshold Positive

	Last 12 Months	
	Pain	
	Number of Episodes	Number of Days
Pain or Soreness Almost Every Day	10	1
	6	2
	4	3
	3	10
	2	30
	1	90

Only disorders reported over the previous 12 months are included: Each section of the spine is considered, and the worker is asked about the presence of almost continuous pain or soreness or painful episodes, according to the distributions indicated.

Other questions concern the number of sick days caused by pain in each individual section of the spine over the last 12 months. The scheme for collecting the anamnestic data is reported in Table 11.3.

To simplify reading the results, the conditions that determine the presence of a positive threshold are highlighted in grey.

For the lower back, the occurrence of acute lumbar pain (Table 11.4) is reported separately; *acute lumbar pain* is defined as "the presence of intense lower back pain with or without irradiation that has caused immobility for at least 2 days or 1 with medication."

In the event of well-recognised and documented back problems, the type and date of the examination and the relevant diagnostic conclusions should be reported (Table 11.5). Of particular interest are:

- Spinal injuries
- Spinal malformations or degenerative diseases
- Herniated disc, and if surgically treated, the date of the procedure

For the upper limbs, the proposed anamnestic model also calls for acquiring information about disorders over the previous 12 months according to a detailed and orderly description of symptoms, the aim being to determine a positive anamnestic threshold and therefore an anamnestic case [Menoni et al., 1996] [De Marco et al., 1996] [Leffert, 1994] based on the following criteria: presence of pain or paraesthesia lasting at least 1 week over the past 12 months or presence of pain or paraesthesia reported at least once a month during the past 12 months.

Table 11.6 illustrates the anamnestic examination scheme for the upper limbs.

TABLE 11.3
Structured Questionnaire Proposed by EPM Research Unit for Reporting Spinal Problems

NB: Mark the figure with the location of pain/soreness and any radiating pain	CERVICAL(soreness, heaviness, pain) ☐ YES ☐ NO			
	SELDOM	AT LEAST 3-4 EPISODES EACH LASTING 2-3 DAYS	AT LEAST 3-4 EPISODES WITH USE OF PAIN MEDICATION OR TREATMENT	ALMOST EVERY DAY
	☐ SORENESS	☐ SORENESS	☐ SORENESS	☐ SORENESS
	☐ PAIN	☐ PAIN	☐ PAIN	☐ PAIN
	RADIATING PAIN	NO UPPER LIMB	RIGHT	LEFT
	NR. OF DAYS OFF WORK DUE TO CERVICAL PROBLEMS ________			
	DORSAL(soreness, heaviness, pain) ☐ YES ☐ NO			
	SELDOM	AT LEAST 3–4 EPISODES EACH LASTING 2-3 DAYS	AT LEAST 3–4 EPISODES WITH USE OF PAIN MEDICATION OR TREATMENT	ALMOST EVERY DAY
	☐ SORENESS	☐ SORENESS	☐ SORENESS	☐ SORENESS
	☐ PAIN	☐ PAIN	☐ PAIN	☐ PAIN
	RADIATING PAIN	NO HEMITHORAX	RIGHT	LEFT
	NR. OF DAYS OFF WORK DUE TO DORSAL PROBLEMS ________			
	LUMBOSACRAL (soreness, heaviness, pain) ☐ YES ☐ NO			
	SELDOM	AT LEAST 3–4 EPISODES EACH LASTING 2-3 DAYS	AT LEAST 3–4 EPISODES WITH USE OF PAIN MEDICATION OR TREATMENT	ALMOST EVERY DAY
	☐ SORENESS	☐ SORENESS	☐ SORENESS	☐ SORENESS
	☐ PAIN	☐ PAIN	☐ PAIN	☐ PAIN
	RADIATING PAIN	NO LOWER LIMB	RIGHT	LEFT
	NR. OF DAYS OFF WORK DUE TO LUMBAR PROBLEMS ________			

TABLE 11.4
Anamnestic Questionnaire for Documenting Acute Lumbago

Acute lumbar pain ☐ Yes ☐ No	
Total number of acute episodes:	
Number of acute episodes in the last year:	☐ Lumbago ☐ Lumbosciatica
Year of first episode:	
Number of days off work due to acute lumbago:	

Note: Acute lumbago refers to episodes of intense pain in the lumbosacral area preventing the sufferer from flexing, bending, or rotating the spine (lower back pain or lumbago). Onset may be acute or insidious and episodes last at least 2 days (or 1 day with pain medication).

TABLE 11.5
Anamnestic Questionnaire for Documenting Known Pathologies

Diagnosis of spinal pathology (already detected):		☐ Yes ☐ No
Herniated lumbosacral disc		
Diagnosed on:	Date: ____________	
Surgically treated:	Date of surgery: ____________	
Pathologies/trauma of the cervical spine	Please specify:	
Pathologies/trauma of the dorsal spine	Please specify:	
Pathologies/trauma of the lumbosacral spine	Please specify:	

11.5 USE OF COLLECTIVE DATA IN PERIODICAL SCREENING PROGRAMS

The proposed screening tools enable subjects to be classified according to the symptoms they reported over the previous 12 months (i.e., positive anamnestic interview).

In the proposed screening model, acute events such as acute lower back pain are specifically reported and the data processed.

Collective data generated by periodical health screening programs, conducted using these tools on groups of exposed workers, can be employed for several different purposes. The main purpose is to detect excessively high percentages of positive cases in the study group, compared to a population of workers with negligible or no occupational exposure.

The program represents:

- A method for evaluating the quality of risk assessment and especially of primary prevention measures
- A resource for planning future primary prevention measures and possibly also for enhancing the quality and quantity of health care surveillance programs

It should be noted that the disorders and pathologies that are the object of such specific health surveillance programs are quite common, and are associated with countless non-work-related factors (including, primarily, gender and age) even among adults who are not exposed to work-related risk.

The occurrence in the general adult population, not exposed to manual load lifting risk, represents the threshold beyond which the higher percentages of disorders reported in exposed groups can be attributed to the specific working conditions on which preventive actions should focus.

To facilitate comparisons, Tables 11.7 and 11.8 list the prevalences, in a group with neglible or no occupational exposure, of workers who are anamnestically positive for cervical, dorsal, and lumbosacral disorders, broken down by gender and age groups, and the annual incidence of acute lower back pain.

TABLE 11.6
Anamnestic Questionnaire for the Upper Extremity, Shoulders, Elbows, and Hands: Questions Designed to Define the Case

SHOULDER PAIN ☐ NO YES ☐	WHEN DID SYMPTOMS BEGIN? (year)	RIGHT	LEFT
RX LX **Regarding reported symptoms, the patient:** ☐ has taken medication **has had:** ☐ physiotherapy ☐ orthopedic/physiatric visit ☐ RX ☐ U.S./MRI	pain during movement		
	pain at rest		
	POSITIVE THRESHOLD ☐ continuous pain ☐ at least 1 week of pain in the last 12 months ☐ at least once a month in the last 12 months		
	MINOR COMPLAINTS Episodes of sub-threshold pain		
NR. OF DAYS OFF WORK DUE TO PAIN OR SORENESS			
ELBOW PAIN ☐ NO YES ☐	**WHEN DID SYMPTOMS BEGIN? (year)**	**RIGHT**	**LEFT**
RX LX **Regarding reported symptoms, the patient:** ☐ has taken medication **has had:** ☐ physiotherapy ☐ orthopedic/physiatric visit ☐ RX ☐ U.S./MRI	pain during movement		
	pain at rest		
	POSITIVE THRESHOLD ☐ continuous pain ☐ at least 1 week of pain in the last 12 months ☐ at least once a month in the last 12 months		
	MINOR COMPLAINTS Episodes of sub-threshold pain		
NR. OF DAYS OFF WORK DUE TO PAIN OR SORENESS			
HAND WRIST PAIN ☐ NO YES ☐	**WHEN DID SYMPTOMS BEGIN? (year)**	**RIGHT**	**LEFT**
SX DX **Regarding reported symptoms, the patient:** ☐ has taken medication **has had:** ☐ physiotherapy ☐ orthopedic/physiatric visit ☐ RX ☐ U.S./MRI ☐ EMG	Pain in grasping		
	Pain in moving		
	Pain during rest		
	Pain at 1° finger		
	Pain at other finger		
	Pain where...		
	Pain where		
	POSITIVE THRESHOLD ☐ continuous pain ☐ at least 1 week of pain in the last 12 months ☐ at least once a month in the last 12 months		
	MINOR COMPLAINTS Episodes of sub-threshold pain		

TABLE 11.7

Results of Anamnestic Screening of the Spine in a Group of Nonexposed Male Subjects (Threshold Positive)

Age	Cervical				Dorsal				Lumbosacral			
	Negative		Positive[a]		Negative		Positive[a]		Negative		Positive[a]	
Group	Number	%	Number	%	Number	%	Number	%	Number	%	Number	%
16–25	45	88.0	6	12.0	50	98.0	1	2.0	48	94.2	3	5.8
26–35	118	82.5	25	17.5	139	97.2	4	2.8	125	87.4	18	12.6
36–45	557	84.3	104	15.7	634	95.3	31	4.7	486	73.5	175	26.5
46–55	430	77.3	126	22.7	534	96.0	22	4.0	346	62.5	208	37.5

Incidence of workers reporting at least one episode of acute lumbago in the last 12 months = 2.3%.

Number of positive cases = 31; Number of negative cases = 1,347; Total = 1,378.

[a] Threshold positive.

TABLE 11.8

Results of Anamnestic Screening of the Spine in a Group of Nonexposed Female Subjects (Threshold Positive)

	Cervical				Dorsal				Lumbosacral			
	Negative		Positive[a]		Negative		Positive[a]		Negative		Positive[a]	
Age Group	Number	%	Number	%	Number	%	Number	%	Number	%	Number	%
16–25	79	93.0	6	7.0	83	96.5	3	3.5	77	90.6	8	9.4
26–35	119	83.2	24	16.8	140	97.9	3	2.1	124	86.7	19	13.3
36–45	146	75.7	47	24.3	183	94.8	10	5.2	145	75.1	48	24.9
46–55	43	58.2	31	41.8	71	96.0	3	4.0	52	70.3	22	29.7

[a] Threshold positive.

It is worth remembering that such comparisons are made with a reference sample population: Domestic data can be used as a term of comparison to detect any major mismatches between the frequency of the symptons (positive threshold) observed in the study group, and the frequency of symptons expected (positive threshold) in the reference group assumed to be un-exposed to work-related risk.

Clearly such comparisons can only be made if the exposed group is large enough for an adequate statistical analysis, and if the subjects have been examined and classified according the methods and criteria proposed in this work.

Naturally the occupational health specialist may use other sources of reference data (i.e., groups of unexposed subjects working in the same company), provided their numbers are sufficiently high.

As for the spine, shoulder disorders can also be compared: Table 11.9 shows the prevalence of anamnestic positive subjects, broken down by gender and age groups, detected in populations not exposed to the risk of biomechanical overload [De Marco et al., 1996].

11.6 CLINICAL-FUNCTIONAL ASSESSMENT: PATHOLOGIES OF INTEREST

11.6.1 PATHOLOGIES OF THE SPINE

Virtually all the pathologies involving the spine (regardless of their aetiology and pathogenetic mechanisms) are of specific interest, even if only for clearing workers to take on specific jobs. However, for the purposes of this work, such pathologies can be roughly broken down into two groups:

- Pathologies not aetiologically correlated to work activities (e.g., pathologies with a constitutional, metabolic, or genetic basis—*prevalently malformations*), but which can be aggravated by biomechanics when present in affected workers.
- Pathologies with multifactorial aetiology in which jobs featuring biomechanical overload conditions may act as the main or one of the main causes. Such forms are centred on *intervertebral disc degeneration processes* as well as acute generic forms.

When jobs such as the manual lifting of heavy loads generate biomechanical overload conditions affecting intervertebral discs, microfractures may appear eventually in the limiting cartilage, as well as microcracks, initially concentric and later radial, in the fibrous ring of the intervertebral disc itself. The consequent degeneration of the intervertebral discs (with loss of fluid and thinning) stretches the longitudinal ligaments, causing bony protuberances around the rim of the vertebra in older individuals and vertebral instability in younger ones (e.g., retro or lateral listhesis). Radial microfissures of the intervertebral disc pave the way for disc herniation. The critical pressure has been estimated on the lumbar intervertebral discs under which no lesions of the vertebral limiting membranes or annulus fibrosus were detected (around 250 kg of compression force on the disc surface): The recommended load

TABLE 11.9

Results of Anamnestic Screening of Upper Extremities in a Nonexposed Group (Threshold Positive): Shoulder

	Positive:[a] Shoulder Females				Positive:[a] Shoulder Males				Sample Breakdown by Gender and Age Group	
	Right		Left		Right		Left		Females	Males
Age Group	Number	%	Number	%	Number	%	Number	%	Number	Number
15–35	1	0.6	0	0	2	1.4	1	0.7	176	139
>35	0	0	0	0	6	3.5	4	2.4	263	171
Total	1	0.2	0	0	8	2.6	5	1.6	439	310

[a] Threshold positive.

TABLE 11.10A
Pathologies Involving the Spine, by Type

Congenital Malformations	Degenerative Disease
Congenital stenosis of cervical medullary canal	Severe lumbar disc disease (disc thinning and limiting somatic dysfunction)
Baastrup syndrome ("kissing spines disease"—development of a neoarthrosis between adjacent spinous processes)	Lumbar disc protrusion with dural sac impingement
Congenital spondylolisthesis due to spondylolysis	Herniated lumbar disc (protruded, contained, migrated)
Scoliosis (at least Cobb 20° and torsion 2)	Outcomes of herniated disc reduction
Scheuermann disease (Schmorl's nodes plus at least 1 wedge vertebra causing a curve of 40°)	Degenerative spondylolisthesis
Sacralisation (fully or partially fused or articulated)	
Klippel-Feil syndrome (vertebral synostosis)	
Newly formed lesions of the bone tissue	

and the lifting index (value = 1) suggested in the RNLE formula tend to ensure this limit is respected.

Lumbar spine pathologies of interest in deciding whether an individual might be allowed to perform manual load (or persons) lifting tasks can be summarised as shown in Table 11.10A and B.

Besides the pathologies indicated above, in terms of the general aims of health care surveillance, it is essential not to overlook general acute forms (recurring lower back pain) that may be an indication of degenerative disease, even if at a very early stage.

Obviously, general acute forms are of interest only for the purpose of judging whether individual sufferers should be allowed to perform certain specific tasks. The same also goes for degenerative disease forms that, in addition, need to be considered on a collective basis since their occurrence may be interpreted (assuming sufficient comparative data are available) as part of a more comprehensive preventative action. Moreover, the decision to allow workers to perform certain tasks should also be based on the condition of other organs and systems (e.g., cardiovascular, respiratory, etc.) and special physiological states (such as pregnancy).

11.6.2 Pathologies of the Shoulder

Clinically, musculoskeletal disorders of the upper limbs, associated with biomechanical overload of the shoulder, feature variable characteristics that are not always easy to differentiate [Colombini et al., 2003]. Such conditions are primarily related to abnormalities of the periarticular soft tissues (tendons and bursae).

Table 11.11 shows the most common diagnoses involving the shoulder (including some nonmedical terms) that can be defined as pathologies due to biomechanical overload of the shoulder.

TABLE 11.10B
Pathologies Involving the Spine, by Severity

Mild Pathologies
Generic dorsal or lumbar spondyloarthropathy with functional deficit (3° SAP according to EPM classification)
Moderate Pathologies
Significant scoliosis (20° Cobb with torsion 2, 30° Cobb with torsion 1+)
Baastrup syndrome
Scheuermann disease (with structured curving of the spine)
Klippel-Feil syndrome (even with only one synostosis)
Cervical or dorsal hernia
Grade 1 spondylolisthesis, spondylolysis
Sacralisation (fully or partially fused or articulated)
Spinal canal stenosis without neurological signs
Severe lumbar disc disease (spondylodiscopathy)
Inverted lumbar lordosis with disc disease
Slight vertebral instability (10/15% in the presence of certain pathologies)
Lumbar protrusion with dural sac impingement
Surgically reduced lumbar disc herniation without adverse outcomes
Severe Pathologies
Herniated disc
Surgically reduced lumbar disc herniation with adverse outcomes
Spinal canal stenosis with root or dural sac impairment
Grade 2 spondylolisthesis (>25% slippage)
Klippel-Feil syndrome (cervical or dorsal synostosis with vertebral instability)
Significant scoliosis (at least Cobb 30° and torsion 2)
Scheuermann disease with approximately 40° structured curving of the spine and lumbar disc disease
Severe vertebral instability (as in spondylolisthesis, Klippel-Feil syndrome, disc disease, fractures with vertebral slippage of 25%)
Degenerative or newly formed lesions of the bones and joints (such as severe osteoporosis, vertebral angioma, etc.)
Systemic disease with severe spinal impairment

The most frequent diagnosis is impingement syndrome (or scapulohumeral arthritis), which may lead to calcific tendinitis or Duplay's periarthritis syndrome, and adhesive pericapsulitis or "frozen shoulder."

This can generally be defined as rotator cuff tendonitis, an inflammation of one or more of the following tendons: supraspinatus, teres minor, subscapularis, and subspinatus.

One of the most widely supported aetiopathogenetic theories suggests that due to repeated mechanical strain, the tendons running through the narrow bony space between the humerus and acromion, tend to thin and tear [Missere, 1998]. Subsequent calcification is caused by insufficient blood flow.

TABLE 11.11
Pathologies Caused by Mechanical Overload of the Shoulder

Tendon Diseases	Synonyms
Supraspinatus tendinopathy, rotator cuff tendinopathy, tendinitis of the long head of the biceps	Impingement syndrome, periarthritis, scapulohumeral
Calcifica tendinitis	Duplay's disease
Acromion deltoid bursitis	Subacromiale bursitis, subdeltoid bursitis
Subspinatus bursitis, subscapular bursitis, bicipital bursitis, subcoracoid bursitis	

TABLE 11.12
Severity of Shoulder Pathologies

Moderate Pathologies	Anatomical Damage
Edema of the supraspinatus tendon	Reversible
Edema of other rotator cuff tendons	Reversible
Edema of the long head of the biceps tendon	Reversible
Shoulder bursitis	Reversible
Mild impingement syndrome	
Severe Pathologies	**Anatomical Damage**
Shoulder bursitis with fibrosis	Irreversible
Tendinosis/fibrosis of the rotator cuff tendons	Irreversible
Calcification of the rotator cuff tendons	Irreversible
Rotator cuff acromionplasty SLAP syndrome	Irreversible

The gliding surfaces of bursae are also prone to inflammation, commonly subacromial bursitis and subdeltoid bursitis.

The subsequent diagnostic process serves to classify the severity of specific shoulder disorders due to biomechanical overload, as described in Table 11.12 with:

Stage 1 (initial): Significant symptoms but instrumental tests negative.
Stage 2 (medium severity): Instrumental testing evidences tendon edema (positive ultrasound signs); damage treatable and reversible.
Stage 3 (severe): Evidence of partial or total rotator cuff lesions or degenerative abnormalities (fibrosis/tendonitis and calcifications) indicates irreversible anatomic damage.

11.7 CLINICAL-FUNCTIONAL ASSESSMENT: ANAMNESIS

The examination model proposed by the EPM Research Unit features the following principal steps in taking the subject's medical history.

11.7.1 Physiological Anamnesis

First, the respondent is required to provide general information: name, address, etc., company name, department, current duties, and length of time in the position. Next, the respondent is asked whether any postural risk factors were present in previous jobs, as defined according to certain basic variants: prolonged sitting, driving vehicles, prolonged standing, manual load lifting, and for the shoulder, repetitive movements with abnormal shoulder postures (e.g., arm raised to shoulder height).

These are in fact the most common postural risk factors reported in the literature as highly significant, if not the main causes of spinal abnormalities.

Previous exposure to a postural risk factor is judged to be significant only if the subject has spent at least 4 years in a given variant, independently of the number of jobs performed.

Length of time in a postural risk means the total number of years that the worker has spent performing that particular task (independently of which company or department).

The physiological anamnesis may consist simply of reporting any sporting activities that may place the spine at risk (such as weight lifting, wind surfing, etc.) or the shoulder (volleyball, weight lifting, etc.), as well as the number and dates of past pregnancies for female workers.

11.7.2 Pathological Anamnesis of Interest

It may be difficult to verify the reliability of reported symptoms especially where workers are assessed for their suitability to take on specific tasks (workers may simulate being suitable or unsuitable). However, it is first necessary to gather information about previous pathologies involving the spine.

Pathologies of particular interest include:

- Scoliosis treated with a brace or corrective exercises
- Herniated disc, and if surgically treated, the date of the procedure
- Number of episodes of radiating or nonradiating acute lumbago, i.e., lower back pain, causing immobility and lasting for at least 2 days or 1 day with medication

The anamnesis also includes reporting/reassessing episodes of lower back pain over the last 12 months (Tables 11.3–11.6, as described above).

At the end of the anamnestic assessment, special attention is focused on the date of onset of the relevant painful symptoms, which may later correlate with specific work-related exposure.

In the event of previously documented back problems, the type and date of the examination and relevant conclusions are reported, as well as a copy of the subject's medical report.

11.8 CLINICAL-FUNCTIONAL ASSESSMENT: PHYSICAL EXAMINATION

The physical examination consists of the following steps.

11.8.1 Spine: Observation of Posture (Figure 11.1)

The exam is performed, if possible, using a scoliosometer to check for the presence of asymmetries.

Any abnormal spinal curvature (lordosis and kyphosis) should be reported, as well as the presence of scoliosis. The presence of a hunch is determined by observing the subject with trunk bent forward, knees extended, arms stretched forward, and hands with palms facing in. The height of the hunch may be measured by placing a ruler over the gibbosity, keeping it perfectly horizontal, and measuring the distance between the ruler and the hemithorax on either side of the centre point.

A hunch of at least 1 cm must be reported. If the subject wore a brace during adolescence to treat significant scoliosis, rather than measure the hunch, the absence of physiological sagittal curves should be assessed (dorsal kyphosis and lumbar lordosis).

With the patient bending forward, the presence of abnormal lumbar-pelvic rhythm should be noted, or the absence of kyphosis in the lumbar lordosis.

If dorsal kyphosis is accentuated, perform the procedure to detect structured curved spine: kyphosis remaining during extension of the dorsal spine (subject sitting, hands behind head, elbows as widely open as possible, spine straight).

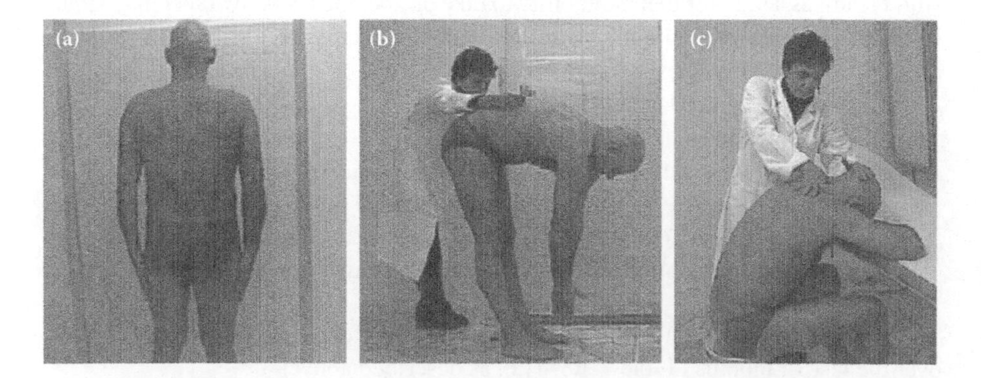

FIGURE 11.1 Observation of the subject and pressure (palpation) manoeuvres: observation of the subject (a), observation of the hunch and lumbar pelvic rhythm (b), and pressure and palpation (c).

11.8.2 Palpation of Paravertebral Musculature, Pressure on Spinous Apophysis, and Intervertebral Spaces (Figure 11.1)

Muscles are palpated along the cervical and dorsolumbar spine: Any pain or paravertebral muscle contractures are detected. At the cervical level, the upper trapezius muscles are palpated at the same time.

During palpation, the subject must relax the neck muscles. The best position is with the subject sitting on a stool, leaning forward, and resting the forearm, hand, and forehead on the bed. The dorso-lumbar muscles are palpated with the subject in the prone position.

The spinous apophyses and intervertebral spaces along the cervical and dorsolumbar spine are palpated in the same positions as described above for the musculature. *Palpation pressure is positive* when pain is reported at at least two apophyses (or intervertebral spaces) or when palpation of muscles in the same area is painful.

11.8.3 Assessment of Analytical Mobility of the Spine

In this test, the subject is asked to perform the principal movements of the segment (maximum unforced range of motion); only pain reported in the mobilised segment of the spine is noted.

- *Movements of the cervical spine: flexion-extension, inclination, rotation* (Figure 11.2): The operator mobilises the subject's head to assess passive mobility. The cervical spine is extended with the subject in the prone position, while for all the other movements the subject is supine. While mobilizing the cervical spine, the operator should keep the subject's shoulder girdle firm. Mobility of the segment is classed as painful when three movements out of six cause pain.
- *Movements of the dorsal, lumbar, and lumbosacral spine: flexion-extension, inclination, rotation* (Figure 11.3): Extension is performed with the subject in the prone position: The subject stretches the spine, leveraging the upper limbs. While extending, the pelvis is kept firm to keep from lifting. During flexion, the subject is seated on the exam bed and asked to flex the trunk forward while the operator keeps the pelvis firm.

  For left and right inclination the subject is seated on the examination table; during inclination the subject must not lift the pelvis that has to remain flat on the table. This is achieved by strapping the contralateral thigh to the table during inclination. During rotation the subject remains seated on the examination table, with upper arms crossed behind the back to keep the shoulder girdle firm. During rotation, the operator holds the knees firmly to keep the pelvis from moving.

During each manoeuvre the operator notes and reports on any pain in the segment of the spine being mobilised.

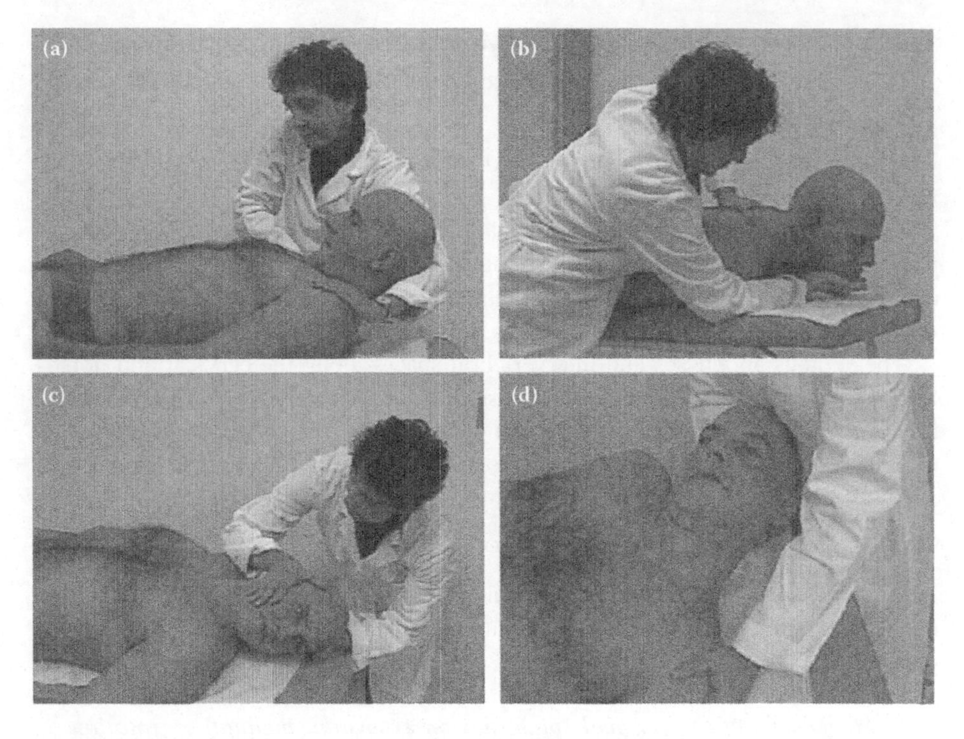

FIGURE 11.2 Examination of cervical mobility: flexion (a), extension (b), rotation (c), and inclination (d).

Mobility of the segment is defined as *pain positive* when the subject reports pain in elective movements (flexion-extension for the lumbar segment and rotation for the dorsal segment) or in at least three movements out of six.

11.8.4 Lasègue/SLR and Wassermann Manoeuvres

This examination schedule also includes a few simple manoeuvres providing additional details about the analytical mobility of the spine, and thus can disclose irritated nerve endings and other significant morphological and functional abnormalities. Besides helping to provide a more comprehensive clinical and functional picture of the spine, preliminary studies have shown that these manoeuvres can steer operators toward suspected pathologies of specific interest. In particular, the Lasègue/SLR (straight leg raising) and Wassermann manoeuvres are performed in the presence of lower back pain radiating down the lower limb reported in the patient history for the last 12 months. During these manoeuvres it may be found that the ischiocrural and ileopsoas muscles are retracted.

11.8.5 Clinical-Functional Assessment: Classification of Results

In order for the results to be processed with a standard method, an original diagnostic classification system has been developed based on clinical and functional

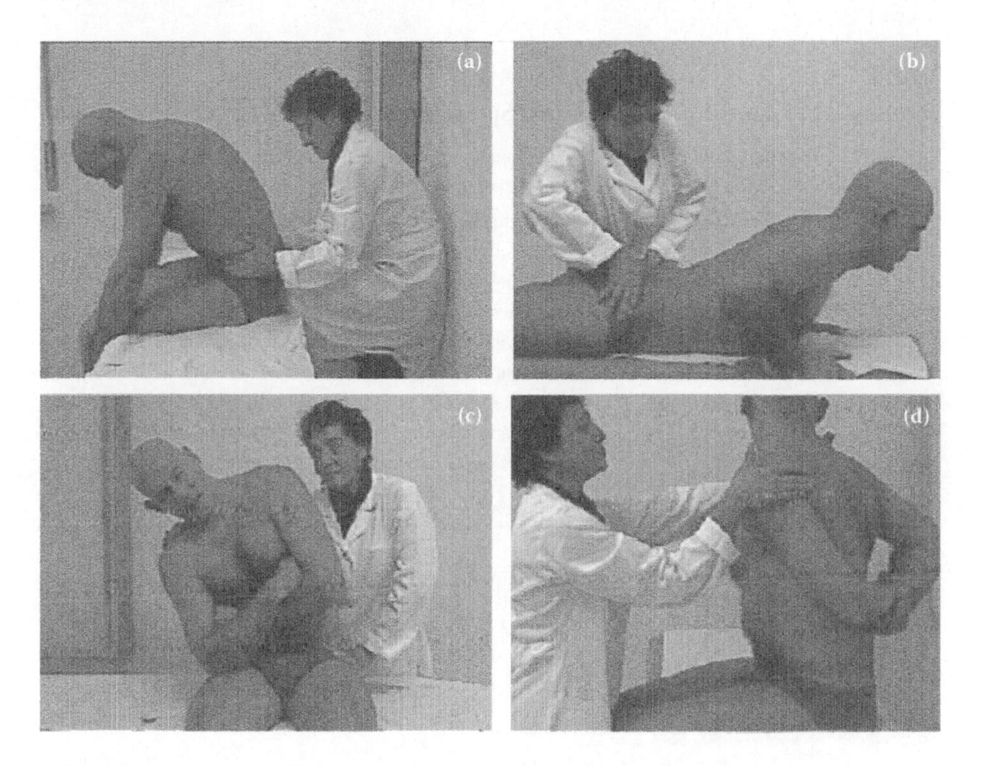

FIGURE 11.3 Examination of dorsolumbar mobility: flexion (a), extension (b), rotation (c), and inclination (d).

elements (Table 11.13). This diagnostic scheme is based on a combination of anamnestic, clinical, morphological, and functional variables, in three separate conditions that may or may not share the same linear course. These conditions of severity are: clinico-functional spondyloarthropathy (SAP) of levels 1, 2, and 3.

In the classification, attention must be paid to the dorsal spine: In terms of clinical and functional disorders, its behaviour differs from that of the other vertebral segments. For instance, symptoms (upper back pain) may commonly appear only later, even when there are radiological changes. This may be due to the decreased mobility of the dorsal segment and the favourable relationship between the anatomical characteristics of the intervertebral foramen and the nerve roots.

11.9 CLINICAL-FUNCTIONAL ASSESSMENT: CLINICAL EXAMINATION OF THE SHOULDER

Shoulder tendonitis should be suspected when the subject reports pain (on moving or lifting loads) during the previous 12 months. Associated pain at rest suggests tendonitis in a more advanced stage. There is no need for a clinical examination if the symptoms are episodic and the subject reports no pain in the last 30 days or more: In this case, if the subject passes the anamnestic threshold, the next examination is scheduled in 6 months' time, or at the next acute episode.

TABLE 11.13
Diagnosis of Clinical and Functional Spondyloarthropathy (SAP) of the Spine

	Spondyloarthropathy Level 1	Spondyloarthropathy Level 2	Spondyloarthropathy Level 3
Cervical	Cervical threshold positive	Cervical threshold positive and pressure (palpation) positive	Cervical threshold positive and pressure (palpation) positive and painful cervical mobility
Dorsal	Dorsal pressure (palpation) positive	Dorsal threshold positive and pressure (palpation) positive	Dorsal threshold positive and pressure (palpation) positive and painful dorsal mobility
Lumbosacral	Cervical threshold positive	Cervical threshold positive and pressure (palpation) positive	Lumbar threshold positive and pressure (palpation) positive and painful lumbar mobility or Lasègue (or Wassermann) positive

A clinical examination is required in the event of:

- Continuous pain
- Current episodic pain
- Pain triggered by a specific cause

Episodes are ongoing when the pain-free interval is less than 30 days; this applies also to pathologies reported subsequently.

The clinical manoeuvres seek to evoke pain both by palpation of specific points or triggers (anterior, lateral, and posterior) and during global movements of the shoulder girdle [Hoppenfeld, 1993] [Fine et al., 1986].

The global movements of the shoulder girdle that the operator must examine, to be actively performed by the subject either sitting or standing, are:

1. *Flexion*: The subject stretches the upper limb forward and up until it is vertical (180° flexion).
2. *Abduction*: The subject performs a complete abduction of the arm, raising it 180° from the frontal plane. The painful arc (between 70 and 120° of abduction) is of particular importance and is regarded as pathognomonic for an impingement syndrome.
3. *External rotation and abduction*: The subject is asked to reach behind his or her head and touch the superior medial angle of the contralateral scapula.
4. *Internal rotation and adduction*: The subject is asked to reach behind his or her back and touch the inferior angle of the controlateral scapula.

Any pain reported while performing the manoeuvres is noted.

To check for the presence of *tendonitis of the long head of the biceps*, the subject is asked to bend the elbow against resistance to the forearm flexed in the supinate position (80–90°): The manoeuvre is positive for pain appearing at the front of the shoulder.

Palpation of the scapolohumeral joint serves to detect the painful anterior trigger, which often indicates the presence of *impingement syndrome*: The subject is sitting or standing, with both arms relaxed; the examiner presses the anterior, lateral, and posterior trigger points of the shoulder.

Shoulder ultrasound is recommended when the exam reveals a painful arc, aching anterior trigger, pain in at least two of the manoeuvres, or pain in the manoeuvre for the long head of the biceps tendon; in most cases, the scan will clarify the diagnosis by ruling out or identifying the inflamed tendons. If an ultrasound cannot be performed, an orthopaedic surgeon or physiatrics should be consulted.

A *shoulder x-ray* is useful only for detecting tendon calcifications (Duplay's disease) or degenerative joint disease.

11.10 INDICATIONS FOR FURTHER CLINICAL INVESTIGATIONS

11.10.1 In-Depth Clinical Investigations: Spine

Tables 11.14 and 11.15 show the main clinical and anamnestic features requiring instrumental testing or a specialist visit.

Subjects with x-ray exams for the relevant segment of the spine, dating back less than 5 years, may be taken into consideration.

11.10.2 Further Clinical Investigations: Shoulder

Radiographic evaluations of the shoulder may be useful for detecting unusual shoulder conformations, or to rule out the outcomes of old traumas.

Ultrasound examinations must be highly targeted and based on an accurate interpretation of clinical signs.

Ultrasound scans must always be bilateral [Missere, 1998] [Colombini et al., 2005].

TABLE 11.14
Main Indications for Instrumental Testing

Reasons for Requesting X-Rays

Lumbosacral RX (2 projections, standing) in the presence of:

- Recurring acute lumbalgia (at least 2 episodes in the last year)
- Radiating lumbar pain, reported subcontinuously for 12 months
- Grade 3 lumbosacral SAP
- Abnormal lumbar-pelvic rhythm and pain on flexion-extension of the spine or marked reduction in lumbar spinal flexion
- Grade 2 lumbar SAP with radiological evidence of lumbar disc disease dating back more than 5 years

Dorsal RX (2 projections, standing) in the presence of:

- Structured curving of the spine
- Grade 3 dorsal SAP

Whole-body RX (2 projections, standing) in the presence of:

- Scoliosis (also measure degrees)

TABLE 11.15
Request for Specialist Visit (Physiatric or Orthopaedic)

- In presence of radiating lumbar pain ask for a TAC or MRI. (The Italian guidelines on herniated discs recommend an MRI 4 to 6 weeks after starting conservative treatment for patients with signs or symptoms of nerve root compression severe enough as to be candidates for surgery. (Istituto Superoire de Sanita, 2008)
- Instrumental testing reporting nonexhaustive or doubtful results.

N.B.

- For patients with a scoliotic hunch and relevant radiographs (performed as shown above) it is necessary to measure the Cobb angle on the radiographs and the degree of vertebral torsion (alternatively, ask for the specialist to measure these values).
- For patients with radiographic evidence of hyperkyphosis, it is necessary to measure the degree of forward curvature on the x-ray and detect any evidence of wedge vertebra.

Occupational health doctors need to know whether a task performed with one arm (either left or right) has caused damage to the side exposed to job-related biomechanical overload, while the other arm not used to perform the same task is normal. For pathologies caused by biomechanical overload of the shoulders due to manual load lifting, both sides should be involved, with the problem sometimes starting in the dominant limb and later affecting the contralateral side.

Ultrasound examinations most often reveal problems with tendons, pinpointing the affected areas and degree of impairment.

Tendonitis is an inflammation of the tendon and tendon sheath characterised by swelling, pain, and loss of function.

Acute tendonitis often follows some sort of trauma, strain or sprain, or overexercise.

Tendonitis is visible at ultrasound with a localised or diffuse increase in the thickness of the relevant anatomical structures, which appear hypoechoic due to the presence of anechoic fluid. This condition is typical of the rotator cuff, where local edema of the nerve fibrils displaces the tendon and creates characteristic images.

Chronic tendinosis is generally caused by repetitive mechanical microtraumas to cross sections of bone, muscle, and tendon held in unnatural or extreme positions or submitted to steady muscle-tendon overload.

When the ultrasound exam is negative and there are positive symptoms, an MRI will be necessary to reach a conclusive diagnosis.

11.11 USE OF COLLECTIVE CLINICAL DATA FOR THE SPINE AND SHOULDER

In order to standardise the results of a complete clinical examination (diagnosis) and evaluate overoccurrence of the pathologies in question, data are provided about their occurrence in the general population. Comparisons with these data will disclose trends for the pathologies in the observed working population.

TABLE 11.16
Occurrence of Certain Spinal Abnormalities in the General Population

Pathology	%	Number	Nationality	Year	Author
Spondylolisthesis	3–6	—	—	1990	Stringa
	2.5	200	United States	1953	Splithoff
	0.6	598 M	United States	1968	Schein
	2.1	4,654 M	United States	1954	Runge
	1.4	21,017	Italy	1990	E.P.M.-Milan
	3	3,000	United States	1950	Allen
Transitional vertebrae	6.5	3,000	United States	1950	Allen
	6.7	4,654	United States	1954	Runge
	1.8	598 M	United States	1968	Schein
	10	200	United States	1953	Splithoff
Complete sacralisation	7.6	200 F	Italy	1990	E.P.M.-Milan
Transverse mega-apophysis	3.3	200 F	Italy	1990	E.P.M.-Milan
Transverse mega-apophysis with pseudoarticulation	2.8	200 F	Italy	1990	E.P.M.-Milan
Schisis	5.9	3,000	United States	1950	Allen
	5	200	United States	1953	Splithoff
	3.1	4,654 M	United States	1954	Runge
	1.7	598 M	United States	1968	Schein
	7.6	200 F	Italy	1990	E.P.M.-Milan
Juvenile osteochondrosis	0.4–8	2:1 M/F 1:1 M/F	—	1978	Rothman
	3.8	210 F	Italy	1990	E.P.M.-Milan
Baastrup syndrome	0.48	21,017	Italy	1990	E.P.M.-Milan
Lumbar hyperlordosis	1.1	21,017	Italy	1990	E.P.M.-Milan
Hyperkyphosis	3.3	210 F	Italy	1990	E.P.M.-Milan

Table 11.16 shows the data for the occurrence of several of the forms listed in the general population [Colombini et al., 1993]. Disagreements among authors may be explained by their use of different clinical diagnostic criteria.

Regarding the prevalence and incidence of herniated discs in the general population, it is difficult to obtain such data broken down by gender and age brackets.

A recent bibliographic source [Istituto Superiore di Sanita, 2008] defines a lifetime prevalence range of 1–3% for lumbar herniated disc in Western countries, which compares well with the male population in the study by Colombini et al. [1993].

Since reference data regarding the prevalence of spinal pathologies are in such short supply (and occasionally even contradictory), it is worth emphasising the importance of detecting and monitoring the anamnestic records of the working population exposed to biomechanical overload risk, as the information often contains alarm bells that may herald worsening situations.

TABLE 11.17
Occurrence of Rotator Cuff Tendonitis in the Finnish General Population

	Rotator Cuff Tendonitis Females		Rotator Cuff Tendonitis Males		Sample Breakdown by Gender and Age Group	
Age Group	Number	%	Number	%	Females	Males
30–39	4	0.6	7	1	609	671
40–49	12	1.7	13	1.9	666	698
50–59	20	3.5	19	3.6	550	526
60–64	0	0	3	5.2	55	56
Total	36	1.9	42	2.1	1,916	1,993

TABLE 11.18
Occurrence of Tendonitis (Shoulder Periarthritis) in a Working Population Not Exposed to the Risk of Biomechanical Overload

	Shoulder Periarthritis Females		Shoulder Periarthritis Males		Sample Breakdown by Gender and Age Group	
Age Group	Number	%	Number	%	Females	Males
15–35	1	0.6	0	0	176	139
>35	7	2.7	1	0.6	263	171
Total	8	1.8	1	0.3	439	310

Tables 11.17 and 11.18 show the occurrence of shoulder tendon disorders broken down by gender and age groups in two populations. The first group (Table 11.17) consists of members of the general Finnish population ($N = 3,909$ subjects) [Miranda et al., 2005] given a specific clinical evaluation. The second group (Table 11.18) derives from the Italian EPM experience.

11.12 SUITABILITY ASSESSMENT

The problem with assessing a worker's suitability to perform specific tasks involving manual lifting is extremely sensitive, for several different reasons.

First, manual lifting jobs are widespread in agriculture, industry, and the service sector, but the weights of the objects lifted and the working environments differ considerably.

Assuming that all jobs must be safe and acceptable for the worker (independently of gender or age), suitability assessments may represent a problem for the occupational health physician, especially when called on to judge whether tasks are acceptable for workers who may have been or are currently suffering from pathologies. It is difficult to decide if a worker should be deemed at risk of developing back problems or exacerbating an existing problem.

Italian official guidelines indicated the maximum weights that individuals (male or female) may lift if suffering from spinal pathologies previously defined as medium or severe. The proposals are referred to here as one of the elements for helping the occupational health doctor to reach decisions regarding relevant cases.

It needs to be emphasised, however, that the *criteria for defining acceptable weights for lifting by subjects with spinal pathologies derive from an empirical assumption.* So far, in fact, no validation studies have been carried out on the criteria to either accept or reject the assumption; hence the criteria need to be used with great caution and pragmatism to test the effectiveness of individual measures to restrict exposure to risk on a case-by-case basis in the field (through close monitoring of individual effects).

The following guidelines have been reconsidered and adapted in light of the information contained in ISO 11228-1 standards concerning the percentage of the general population protected from various weight constants. Apart from the validation aspects, the guidelines are useful insofar as they form the basis for *synthetic indexes of specific risk exposure* (even for subjects with spinal pathologies). On the one hand they allow classifying, in terms of risk exposure, the different activities performed, and on the other, to highlight criticalities and therefore allow classifying priorities for organisational, structural, and educational interventions so as to contain exposure levels.

The lifting index for hypersusceptible subjects (i.e., those with pathologies) is calculated by combining the procedures included in ISO 11228-1 and EN 1005-2 for computing the lifting index via the RNLE with the information (Annex C to the regulations) on reference weight constants, the aim being to use the methodologies presented in this volume to generate a lifting index (LI) based on weight constants defined as follows:

- For males with pathologies of medium severity the index is computed taking 15 kg as the weight constant.
- For females with medium to severe pathologies and males with severe pathologies the index should be recalculated with a weight constant of 10 kg.

Subjects could be judged suitable (with limitations) when the lifting index, thus specifically computed, is less than or equal to 1. However, it should be emphasised that loads must only be lifted vertically between the height of the knees and the shoulders; in more complex cases, the amount of time assigned in the shift to manual lifting must be limited.

In any case, the workers require close clinical monitoring and there needs to be an undertaking to seek constant improvements in load lifting conditions for these subjects.

The problem of choosing the correct reference values, as evidenced in relation to manual load lifting tasks, also applies to pushing and pulling tasks. Table 11.19 provides values in percentage terms for acceptable initial and continuing forces applicable to different population percentiles (90th, 99th, 99.9th) during pushing or pulling tasks at a frequency of 1 action every 5 min, over distances of 7 m and forces applied at a height from the floor of 90–95 cm, for both males and females [Snook and Ciriello, 1991]. The upper part of the table indicates the reference values that

TABLE 11.19
Pushing and Pulling Tasks: Percentiles of Reference Force (kg), for 1 Action every 5 min, over a Distance of 7.5 m, with a Pushing Height of around 90–95 cm from the Ground

	Values for the 90th Percentile of the Healthy Population		Values for the 99th Percentile of the Population		Values for the 99.9th Percentile of the Population	
Pushing Tasks	**Initial Force**	**Sustained Force**	**Initial Force**	**Sustained Force**	**Initial Force**	**Sustained Force**
Males	25	15	20	9	13	4
			–20%	–40%	–48%	–73%
Females	19	9	14	5	10	—
			–26%	–44%	–47%	
Pushing Tasks	**Initial Force**	**Sustained Force**	**Initial Force**	**Sustained Force**	**Initial Force**	**Sustained Force**
Males	24	16	15	8	8	3
			–37%	–50%	–66%	–80%
Females	19	10	13	4	8	—
			–31%	–40%	–68%	

Source: Data adapted from Snook, S.H., and Ciriello, V.M., *Ergonomics* 34(9): 1197–1213, 1991.

would protect 90% of the population. The reference values for subjects with back problems are:

- For subjects with back problems of medium severity the values are those that protect 99% of the general population.
- For subjects with severe back problems the values are those that protect 99.9% of the general population.
- For all subjects with back problems, pushing and pulling tasks should not cover distances of over 7.5–15 m (unless there is a stop along the way), with no more than one push a minute.

Table 11.19 shows that in pushing tasks, the values for the 99th percentile feature a starting push that is 20–25% below the values suggested for the healthy population; for the maintenance of the push the value is 40–45% lower. In pulling tasks the starting peak is 30–35% lower; for the maintenance part of the pulling task it is 40–50% lower.

With regard to the values referring to the 99.9th percentile of the population, the starting push is 45–48% lower than for the healthy population, and the maintenance of the push is 70% lower. For pulling tasks, the starting pull is 65–68% lower and the maintenance of the pull is 80% lower. Subjects with spondylolistesis or vertebral instability, the severity of which may place them in the medium group, should be treated as severe in respect to pushing and pulling tasks. Pushing and pulling actions generate major shear stress on the vertebrae; therefore these kinds of problems may be particularly aggravated by such tasks. Therefore, the reference values that are safe for 99.9% of the population must be applied to subjects with spondylolistesis or vertebral instability.

In brief, it can be stated that in reassigning tasks to workers with back problems, the whole system of combining jobs with pathological cases reflects a constantly changing landscape. In other words, every effort must be made to continue improving both jobs and working conditions insofar as the involvement of the spine is concerned: Accordingly, the tasks performed by workers with back problems need to be periodically analysed (while never losing sight of the need to safeguard their acquired professional skills).

The aim is obviously that of more effectively protecting individuals whose particular characteristics render them more susceptible to specific work-related risks, as overall working conditions are gradually improved.

Wherever exposure to manual lifting risk has been clearly assessed through accurate job analysis, a list can be drawn up of jobs that can be assigned to workers with back problems.

However, besides adopting clearly defined, approved, and technically acceptable management criteria (e.g., meticulous risk analysis for the various tasks and appropriate diagnostic criteria), it is also essential to put together dedicated databases and provide the necessary informatic support.

It is equally important to deal with the numerous other aspects that may emerge when qualified workers are reassigned to new or modified tasks (i.e., psychological and motivational issues, education and training, remuneration, etc.). However, these matters are not the focus of this chapter.

The model for managing the assignment of workers with back problems to suitable jobs requires close cooperation between the occupational medicine physician and the prevention staff or other bodies responsible for assessing and planning corrective interventions. Such interventions are ongoing and as such depend on many other decision makers (e.g., employers, HR departments, etc.), who cannot and must not feel excluded from the management of this issue.

References

ACGIH. 2005. *Lifting TLV® physical agents*. 7th ed., Publication 7DOC-734. Cincinnati.

Boda S.V., Bhoyar P., Garg A. 2010. Validation of revised NIOSH lifting equation and 3D SSP model to predict risk of work-related low back pain. In *Proceedings of the Human Factors and Ergonomics Society 54th Annual Meeting 2010*, vol. 5, pp. 1185–1189.

CEN. 2001. EN 1005-1. *Safety of machinery—Human physical performance—Part 1: Terms and definitions*.

CEN. 2002. EN 1005-3. *Safety of machinery—Human physical performance—Part 3: Recommended force limits for machinery operation*.

CEN. 2003. EN 1005-2. *Safety of machinery—Human physical performance—Part 2: Manual handling of machinery and component parts of machinery*.

CEN. 2004. EN 1005-4. *Safety of machinery—Human physical performance—Part 4: Evaluation of working postures and movements in relation to machinery*.

CEN. 2007. EN 1005-5. *Safety of machinery—Human physical performance—Part 5: Risk assessment for repetitive handling at high frequency*.

Colombini D., Menoni O., Occhipinti E., et al. 2005. Criteria for classification of upper limb work-related musculo-skeletal disorders due to biomechanical overload in occupational health. Consensus document by an Italian Working Group. *Med Lav* 96(2): 5–24

Colombini D., Occhipinti E., Alvarez-Casado E., Hernandez-Soto A., Waters T.R. 2009. Procedures for collecting and organizing data useful for the analysis of variable lifting tasks and for computing the VLI. Paper presented at Proceedings of the 17th Triennial Congress of the International Ergonomics Association, Beijing, China, August 9–14.

Colombini D., Occhipinti E., Cairoli S., et al. 2003. Musculoskeletal conditions of the upper and lower limbs as an occupational disease: What kind and under what conditions. Consensus document of a national working-group. *Med Lav* 94(3): 312–329.

Colombini D., Occhipinti E., Menoni O., Bonaiuti D., Grieco A. 1983. Bad work posture and the pathology of the locomotor system. *Med Lav* 74(3): 198–210.

Colombini D., Occhipinti E., Menoni O., et al. 1993. Diseases of the dorsal-lumbar spine and manual handling of loads: Guidelines for fitness assessment. *Med Lav* 84(5): 416–432.

Council of European Communities. 1990. Council directive of 29 May 1990 on the minimum health and safety requirements for the manual handling of loads where there is a risk particularly of back injury to workers (fourth individual directive within the meaning of Article 16 (1) of Directive 89/391/EEC) (90/269/EEC). *Off J Eur Comm* 156/9. http://eur-lex.europa.eu/LexUriServ/LexUriServ.do?uri=OJ:L:1990:156:0009:0013:EN:PDF (accessed March 13, 2012).

De Marco F., Menoni O., Ricci M.G., Bonaiuti D., Colombini D., Occhipinti E. 1996. Studi clinici in popolazioni lavorative: Valore e significato dei rilievi anamnestici, dei test clinici e degli esami strumentali per la diagnosi dei WMSDs. *Med Lav* 87(6): 561–580.

Dempsey P.G. 1999. Utilizing criteria for assessing multiple-task manual materials handling jobs. *Int J Ind Ergonomics* 24: 405–416.

Dempsey P.G. 2002. Usability of the revised NIOSH lifting equation. *Ergonomics* 45(12): 817–828.

Dempsey P.G., Burdorf A., Fathallah F., Sorock G.S., Hashemi L. 2001. Influence of measurement accuracy on the application of the 1991 NIOSH equation. *Appl Ergonomics* 32: 91–99.

Dempsey P.G., Fathallah F. 1999. Application issues and theoretical concerns regarding the 1991 NIOSH equation asymmetry multiplier. *Int J Indust Ergonomics* 23: 181–191.

Elfeituri F.E., Taboun S.M. 2002. An evaluation of the NIOSH lifting equation: A psychophysical and biomechanical investigation. *Int J Occup Safety Ergonomics* 8(2): 243–258.

European Agency for Safety and Health at Work. 2008. Work-related musculoskeletal disorders: Prevention report.

European Foundation (European Foundation for the Improvement of Living and Working Conditions). 2007. Fourth European Working Conditions Survey. *Office for Official Publications of the European Communities*. Loughlinstown, Dublin, Ireland. http://www.eurofound.europa.eu/pubdocs/2006/98/en/2/ef0698en.pdf (accessed March 13, 2012).

Fine L.J., Silverstein B.A., Armstrong T.J., Anderson C.A., Sugano D.S. 1986. Detection of cumulative trauma disorders of upper extremities in the workplace. *J Occup Med* 28(8): 674–678.

Garg A. 1976. A metabolic rate prediction model for manual materials handling jobs. PhD. dissertation, University of Michigan.

Grieco A., Occhipinti E., Colombini D., Molteni G. 1997. Manual handling of loads: the point of view of experts involved in the application of EC Directive 90/269. *Ergonomics* 40(10): 1035–1056.

Harkness E.F., Macfarlane G.J., Nahit E.S., Silman A.J., Mcbeth J. 2003. Mechanical and psychosocial factors predict new onset shoulder pain: A prospective cohort study of newly employed workers. *Occup Environ Med* 60(11): 850–857.

Honsa K., Vennettilli M., Mott N., Silvera D., Niechwiej E., Wagar S. 1998. The efficacy of the NIOSH (1991) hand-to-container coupling factor. In *Proceedings of the 30th Annual Conference of the Human Factors Association of Canada*, pp. 253–258.

Hoozemans M.J., Van der Beek A.J., Fring-Dresen M.H., Van der Woude L.H., Van Dijk F.J. 2002. Low-back and shoulder complaints among workers with pushing and pulling tasks. *Scand J Work Environ Health* 28(5): 293–303.

Hoppenfeld S. 1993. *L'esame obbiettivo dell'apparato locomotore*. Aulo Gaggi Editore, Bologna, Italy.

HSE (Health and Safety Executive). 2004. *Manual handling. Manual handling operations regulations 1992* (as amended). *Guidance on Regulations L23*. 3rd ed. HSE Books, London.

ILO (International Labour Organisation). 1967a. Maximum weight convention (N° 127). Geneva: ILO. http://www.ilo.org/ilolex/cgi-lex/convde.pl?R128 (accessed July 15, 2011).

ILO (International Labour Organisation). 1967b. Maximum weight recommendation (N° 128). ILO, Geneva. http://www.ilo.org/ilolex/cgi-lex/convde.pl?C127 (accessed July 15, 2011).

ILO (International Labour Organisation). 1988. *Maximum weights in load lifting and carrying*. Occupational Safety and Health Series 59. http://www.ilo.org/public/libdoc/ilo/1988/88B09_19_engl.pdf (accessed July 15, 2011).

ILOLEX. 2011. Ratifications of Convention No. C127. http://www.ilo.org/ilolex/cgi-lex/ratifce.pl?C127 (accessed July 15, 2011).

INAIL. 2009. Rapporto Annuale Analisi dell'andamento infortunistico 2008. Milano. http://www.lavoro.gov.it/NR/rdonlyres/3F0FBA56-72D6-4F17-9A5B-922940753881/0/RappAnnuale2008.pdf (accessed March 12, 2012).

ILO-CIS (International Occupational Safety and Health Information Centre). 1962. Manual lifting and carrying. *Information Sheet No. 3,* Geneva.

ISO. 2000. ISO 11226. *Ergonomics—Evaluation of static working postures.*

ISO. 2003. ISO 11228-1. *Ergonomics—Manual handling—Lifting and carrying.*

ISO. 2007a. ISO 11228-2. *Ergonomics—Manual handling—Pushing and pulling.*

ISO. 2007b. ISO 11228-3. *Ergonomics—Manual handling—Handling of low loads at high frequency.*

Istituto Superiore di Sanita. 2008. Programma nazionale per le linee guida. Appropriatezza della diagnosi e del trattamento chirurgico dell'ernia del disco lombare sintomatica. Doc n° 9 2005 (rev. 2008). http://www.snlg-iss.it/cms/files/LG_ernia_disco.pdf (accessed March 12, 2012).

Jäger M., Luttmann A. 1999. Critical survey on the biomechanical criterion in the NIOSH method for the design and evaluation of manual lifting tasks. *Int J Indust Ergonomics* 23(4): 331–337.

Lavender S.A., Li Y.C., Natarajan R.N., Andersson G.B.J. 2009. Does the asymmetry multiplier in the 1991 NIOSH lifting equation adequately control the biomechanical loading of the spine? *Ergonomics* 52(1): 71–79.

Leffert R.D. 1994. Thoracic outlet syndrome. *Am Acad Orthop Surg* 2(6): 317–325.

Liberty Mutual. 2004. Manual materials handling guidelines. http://libertymmhtables.libertymutual.com/CM_LMTablesWeb/pdf/LibertyMutualTables.pdf (accessed July 15, 2011).

Maiti R., Bagchi T.P. 2006. Effect of different multipliers and their interactions during manual lifting operations. *Int J Indust Ergonomics* 36(11): 991–1004.

Marras W.S., Fine L.J., Ferguson S.A., Waters T.R. 1999. The effectiveness of commonly used lifting assessment methods to identify industrial jobs associated with elevated risk of low-back disorders. *Ergonomics* 42(1): 229–245.

Menoni O., De Marco F., Colombini D., Occhipinti E., Vimercati C., Panciera D. 1996. Studi clinici in popolazioni lavorative: Un modello per l'indagine anamnestica delle patologie degli arti superiori e sue modalità applicative. *Med Lav* 87(6): 549–560.

Miranda H., Viikari-Juntura E., Heistaro S., Heliövaara M., Riihimäki H.A. 2005. Population study on differences in the determinants of a specific shoulder disorder versus nonspecific shoulder pain without clinical findings. *Am J Epidemiol* 161(9): 847–855.

Missere M. 1998. *L'ecografia dell'arto superiore. Le patologie muscolo-tendinee occupazionali: Il percorso diagnostico con la collaborazione del reumatologo e del medico legale.* Athena Audiovisuals, Modena, Italy.

Mital A., Nicholson A.S., Ayoub M.M. 1993. *A guide to manual materials handling*. Taylor & Francis, London.

Mital A., Ramakrishnan A. 1999. A comparison of literature-based design recommendations and experimental capability data for a complex manual materials handling activity. *Int J Indust Ergonomics* 24(1): 73–80.

Monnington S.C, Pinder A.D.J., Quarrie C. 2002. *Development of an inspection tool for manual handling risk assessment.* HSL/2002/30 Crown copyright. Sheffield. http://www.hse.gov.uk/research/hsl_pdf/2002/hsl02-30.pdf (accessed July 15, 2011).

National Technical Information Service. 1991. *Scientific support documentation for the revised 1991 NIOSH lifting equation*. PB91-226274. Springfield, VA.

NIOSH. 1981. *Work practices guide for manual lifting.* NIOSH technical report. Publication N° 81-122. U.S. Department of Health and Human Services.

Occhipinti E., Colombini D., Grieco A. 1993. Study of distribution and characteristics of spinal disorders using a validated questionnaire in a group of male subjects not exposed to occupational spinal risk factors. *Spine* 18(9): 1150–1159.

Occhipinti E., Colombini D., Menoni O., Grieco A. 1985. Changes of the spine in working populations. 1. Data on a male control group. *Med Lav* 76(5): 387–398.

Occhipinti E., Colombini D., Molteni G., Menoni O., Boccardi S., Grieco A. 1988. Development and validation of a questionnaire in the study of spinal changes in a working population. *Med Lav* 79(5): 390–402.

Potvin J.R., Bent L.R. 1997. NIOSH equation horizontal distances associated with the Liberty Mutual (Snook) lifting table box widths. *Ergonomics* 40(6): 650–655.

Putz-Anderson V., Waters T.R. 1991. Revisions in NIOSH *Guide to Manual Lifting*. Paper presented at National Strategy for Occupational Musculoskeletal Injury Prevention—Implementation Issues and Research Needs, University of Michigan.

Ribeiro M.L.L., Remor E. 2009. Proposed procedures for measuring the lifting task variables required by the revised NIOSH lifting equation—A case study. *Int J Indust Ergonomics* 39(1): 15–22.

Saleem J.J., Kleiner B.M., Nussbaum M.A. 2003. Empirical evaluation of training and a work analysis tool for participatory ergonomics. *Int J Indust Ergonomics* 31(6): 387–396.

Silverstein B.A., Bao S.S., Fan Z.J., et al. 2008. Rotator cuff syndrome: Personal, work-related psychosocial and physical load factors. *J Occup Environ Med* 50(9): 1062–1076.

Snook S.H., Ciriello V.M. 1991. The design of manual handling tasks: Revised tables of maximum acceptable weights and forces. *Ergonomics* 34(9): 1197–1213.

Washington State Department of Labour and Industries. 2008. Hazard zone jobs checklist. http://www.lni.wa.gov/wisha/ergo/evaltools/hazardzonechecklist.pdf (accessed July 15, 2011).

Waters T.R. 1991. Strategies for assessing multi-task manual lifting jobs. Paper presented at Proceedings of the Human Factors Society 35th Annual Meeting, San Francisco, CA.

Waters T.R. 2006. Revised NIOSH lifting equation. In *The occupational ergonomics handbook: Fundamentals and assessment tools for occupational ergonomics*, ed. W. Marras, W. Karwowski, 46-1/28. 2nd ed. CRC Press, Boca Raton, FL.

Waters T.R. 2007. When is it safe to manually lift a patient? *Am J Nursing* 107: 53–58.

Waters T.R., Baron S.L., Kemmlert K. 1998. Accuracy of measurements for the revised NIOSH lifting equation. *Appl Ergonomics* 29(6): 433–438.

Waters T.R., Baron S.L., Piacitelli L.A., et al. 1999. Evaluation of the revised NIOSH lifting equation. A cross-sectional epidemiologic study. *Spine* 24(4): 386–395.

Waters T.R., Lu M.L., Occhipinti E. 2007. New procedure for assessing sequential manual lifting jobs using the revised NIOSH lifting equation. *Ergonomics* 50(11): 1761–1770.

Waters T.R., Lu M.L., Piacitelli L.A., Werren D., Deddens J.A. 2011. Efficacy of the revised NIOSH lifting equation to predict risk of low back pain due to manual lifting: Expanded cross-sectional analysis. *J Occup Environ Med* 53(9): 1061–1067.

Waters T.R., Occhipinti E., Colombini D., Alvarez-Casado E., Hernandez A. 2009. The variable lifting index: A tool for assessing manual lifting tasks with highly variable task characteristics. Paper presented at Proceedings of the 17th Triennial Congress of the International Ergonomics Association, Beijing, August 9–14.

Waters T.R., Putz-Anderson V., Garg A. 1994. *Applications manual for the revised NIOSH lifting equation.* DHHS (NIOSH) Publication 94-110. National Institute for Occupational Safety and Health, Centers for Disease Control and Prevention.

Waters T.R., Putz-Anderson V., Garg A., Fine L.J. 1993. Revised NIOSH equation for the design and evaluation of manual lifting tasks. *Ergonomics* 36(7): 749–776.

Index

D

E

F

G

H

O

P

Q

R

S

W

X

Y